T0330018

Innovative Governance Models for Emerging Technologies

Innovative Governance Models for Emerging Technologies

Edited by

Gary E. Marchant
Kenneth W. Abbott
Braden Allenby

Arizona State University, US

Edward Elgar

Cheltenham, UK • Northampton, MA, USA

Published by
Edward Elgar Publishing Limited
The Lypiatts
15 Lansdown Road
Cheltenham
Glos GL50 2JA
UK

Edward Elgar Publishing, Inc.
William Pratt House
9 Dewey Court
Northampton
Massachusetts 01060
USA

A catalogue record for this book
is available from the British Library

Library of Congress Control Number: 2013943235

This book is available electronically in the ElgarOnline.com Law Subject Collection, E-ISBN 978 1 78254 564 4

ISBN 978 1 78254 563 7

Typeset by Columns Design XML Ltd, Reading
Printed and bound in Great Britain by T.J. International Ltd, Padstow

Contents

v

Contributors

Joshua W. Abbott, Executive Director, Center for Law, Science & Innovation, Sandra Day O'Connor College of Law, Arizona State University, US

Kenneth W. Abbott, Jack E. Brown Professor of Law, and Faculty Fellow of the Center for Law, Science & Innovation at the Sandra Day O'Connor College of Law at Arizona State University, and Professor of Global Studies, School of Politics & Global Studies, Arizona State University, US

Braden Allenby, President's Professor and Lincoln Professor of Engineering and Ethics, School of Sustainable Engineering and the Built Environment, Ira A. Fulton Schools of Engineering, and Director, Center for Earth Systems Engineering and Management, Arizona State University, US

Michael Baram, Professor of Law Emeritus, Boston University School of Law, US

Diana M. Bowman, Assistant Professor, Health Management and Policy and the Risk Science Center, School of Public Health, University of Michigan, US

Jennifer Kuzma, Professor, School of Public and International Affairs and Goodnight-Glaxo Wellcome Distinguished Professor in the Social Sciences, and Co-Director of the Genetic Engineering and Society Center, North Carolina State University, US

Preben H. Lindøe, Professor, University of Stavanger, Norway

Rachel A. Lindor, Research Director, Center for Law, Science & Innovation, Sandra Day O'Connor College of Law, Arizona State University, and M.D. candidate, Mayo Medical School, US

Timothy F. Malloy, Professor of Law and Faculty Director, Sustainable Technology & Policy Program, University of California at Los Angeles School of Law, US

Gregory N. Mandel, Associate Dean for Research and Peter J. Liacouras Professor of Law, Beasley School of Law, Temple University, US

Gary E. Marchant, Regents' Professor of Law, Lincoln Professor of Emerging Technologies, Law and Ethics, and Faculty Director and Faculty Fellow of the Center for Law, Science & Innovation at the Sandra Day O'Connor College of Law, Arizona State University, US

Molly Masterton, J.D. candidate, The George Washington University Law School

LeRoy Paddock, Associate Dean for Environmental Law Studies, The George Washington University Law School, US

John Paterson, Professor, School of Law, University of Aberdeen, UK

Marc A. Saner, Director and Associate Professor, Institute for Science, Society and Policy, University of Ottawa, Canada

Wendell Wallach, Lecturer and Research Group Chair, Interdisciplinary Center for Bioethics, Yale University, US

Acknowledgment

This book is the work product of a grant (SES 0921806) from the National Science Foundation entitled "Adapting Law to Rapid Technological Change." In addition, four of the chapters (Paddock, Saner, Kuzma, Baram) are based on papers originally presented at a workshop funded by a grant from the Department of Energy Office of Science (grant # 0013858). The views represented in the volume are those of the respective authors and do not represent the views of the National Science Foundation, the Department of Energy, or the US Government.

1. Introduction: the challenges of oversight for emerging technologies

Kenneth W. Abbott

1.1 INTRODUCTION

We are in the midst of one of the greatest periods of scientific and technological innovation in human history. Scientific discoveries and technological developments are fueling explosive advances in biotechnology, personalized medicine and synthetic biology; applied neuroscience; nanotechnology; information technology and artificial intelligence; robotics; geoengineering and other fields. Each of these emerging technologies promises almost unfathomable social and personal benefits. As a result, governments are actively promoting them, while researchers and the private sector are devoting enormous resources to their development. The result has been rapid and widespread commercialization, production and application. Yet each of these technologies also carries the possibility of significant risks to health, safety and the environment. And each entails other potential impacts that raise broader social, economic and ethical concerns.

To be sure, many of the issues raised by the current emerging technologies are generic: they attend every significant new technology. As Christopher Bosso argues, society must balance the potential benefits of every technological innovation against its potential risks and potential economic, social and personal impacts. Benefits, moreover, typically appear near-term and tangible – in the light of self-interest as well as natural optimism – while risks appear distant and intangible.[1]

Nonetheless, the current suite of emerging technologies poses challenges for regulatory oversight that are quantitatively, if not qualitatively,

---

[1] Christopher Bosso, "The Enduring Embrace: The Regulatory *Ancien Régime* and Governance of Nanomaterials in the U.S.", 9 NANOTECHNOLOGY LAW & BUSINESS 4 (June 2013).

greater than those posed by most earlier innovations.[2] These challenges stem from the breadth and power of individual platforms such as biotechnology, nanotechnology and synthetic biology. They also derive from the convergence of multiple technologies, such as the application of advanced information processing to enhance biotechnology or robotics, or the use of synthetic biological forms as "production facilities" for nanoscale fabrication,[3] producing even more powerful hybrids. Finally, the simultaneous development and introduction of so many complex technological platforms creates significant difficulties for oversight.

This volume considers the challenges for regulatory oversight posed by current emerging technologies, and proposes innovative models of governance for responding to those challenges. The governance innovations suggested by the contributing authors involve both modifications of traditional regulatory procedures and "governance" approaches to oversight that go beyond traditional regulation. As such, the volume makes a significant contribution to the evolving debate on oversight of emerging technologies.

This introductory chapter outlines and analyzes some of the major challenges to regulatory oversight, setting the stage for consideration of potential responses. To structure the analysis, the chapter uses as a point of reference the risk governance framework and related materials prepared by the International Risk Governance Council (IRGC).[4] While this volume focuses primarily on the regulatory system of the United States, the challenges discussed here and in the work of the IRGC arise in all countries affected by emerging technologies. Some chapters in the volume, moreover, expressly consider developments in other countries.[5] And many of the proposals advanced by the contributing authors are broadly applicable – at least in principle – across diverse economic and regulatory systems.

---

[2]   Braden Allenby, "The Dynamics of Emerging Technology Systems" (this volume, Chapter 2).

[3]   LeRoy Paddock and Molly Masterton, "An Integrated Framework for Governing Emerging Technologies such as Nanotechnology and Synthetic Biology" (this volume, Chapter 4).

[4]   IRGC, An Introduction to the IRGC Risk Governance Framework (2008), available at http://www.irgc.org/risk-governance/irgc-risk-governance-framework (accessed July 7, 2013).

[5]   For example, Diana M. Bowman, "The Hare and the Tortoise: An Australian Perspective on Regulating New Technologies and their Products and Processes" (this volume, Chapter 8); Preben H. Lindøe, Michael Baram and John Paterson, "Reliance on Industrial Standards to Prevent Major Accidents in Offshore Oil and Gas Operations" (this volume, Chapter 12).

1.2 CHALLENGES TO REGULATORY OVERSIGHT

1.2.1 Pacing

One of the major challenges posed by emerging technologies is the "pacing problem:" how can regulatory oversight arrangements keep pace with rapid scientific and technological innovation?[6] To be sure, technological innovation is a complex process, and not necessarily a rapid one. As Timothy F. Malloy outlines in Chapter 6, innovation can be seen as involving six stages: recognition of a need or opportunity; basic research; applied research; development; commercialization (including production of products and applications of technology); and diffusion, leading to widespread adoption.[7] Frequently, then, the innovation process spans substantial time periods. Current emerging technologies, however, are moving through these stages at an extremely rapid pace, "in many cases on exponential or near-exponential paths."[8]

This pace is driven primarily by private economic incentives and first-mover advantages. But the process of innovation is accelerated by a range of governmental "push" and "pull" incentives,[9] motivated by perceived social benefits and a widespread innovation culture that views innovation as "the key ingredient of any effort to improve people's quality of life."[10] The result is a "tortoise and hare" problem:[11] technological innovation – the hare – is inherently fast and is further

---

[6] See The Growing Gap Between Emerging Technologies and Legal-Ethical Oversight (Gary E. Marchant, Braden R. Allenby and Joseph R. Herkert, eds, 2011).

[7] Timothy F. Malloy, "Integrating Technology Assessment into Government Technology Policy" (this volume, Chapter 6). Malloy adopts the six-stage process introduced in Everett M. Rogers, DIFFUSION OF INNOVATIONS 132–50 (4th ed. 1995).

[8] Gary E. Marchant and Wendell Wallach, "Governing the Governance of Emerging Technologies" (this volume, Chapter 7).

[9] Malloy, supra note 7; Gregory F. Nemet, "Demand-Pull, Technology-Push, and Government-Led Incentives for Non-Incremental Technical Change," 5 RESEARCH POLICY 700, 702 (2009).

[10] OECD, The OECD Innovation Strategy: Getting a Head Start on Tomorrow, available at http://www.oecd.org/site/innovationstrategy/#d.en.192273 (accessed July 7, 2013).

[11] Jennifer Kuzma, "Properly Paced or Problematic?: Examining the Past and Present Governance of GMOs in the United States" (this volume, Chapter 9); Bowman, supra note 5.

accelerated by market and public incentives; governmental oversight – the tortoise – is by comparison inherently slow.

Consider the ideal oversight process formulated by the IRGC, designed "to enable societies to benefit from change while minimizing the negative consequences of the associated risks."[12] The IRGC risk governance framework encompasses five interrelated phases that expand on the traditional categories of risk assessment, management and communication. "Pre-assessment" involves early warning of potential risks, framing those risks in terms of their technical characteristics and the issues perceived by stakeholders and society, and identifying potentially applicable analytical tools and regulatory frameworks. "Appraisal" comprises both a scientific risk assessment and a "concern assessment," which analyzes stakeholder, social and cultural perceptions of potential risks. "Characterization" entails determining whether the likely risks are acceptable and identifying appropriate responses. "Risk management" includes identifying, evaluating and implementing measures to avoid, reduce or transfer risks. And "communication" is a two-way process, with oversight bodies informing stakeholders and society about risks and providing them with opportunities for voice throughout the process. Importantly, the IRGC framework goes well beyond scientific and material analysis of risks (for example, their measurable qualities and the probability of their occurrence) to incorporate social and psychological considerations in every phase. Thus, for example, assessment should consider the broader risk culture, as well as the ways in which different groups perceive risks and benefits "when values and emotions come into play."[13]

This process, done properly, clearly requires substantial amounts of time. Organizing oversight around five stages makes the process inherently lengthy. Framing and appraisal require oversight bodies to gather and analyze extensive scientific and technical information. For some emerging technologies, it will be necessary to generate new data, and even to develop new testing methods and assessment models.[14] Framing and appraisal also require oversight bodies to identify stakeholders and potentially concerned social groups, as well as representatives of the public at large, and to gather and assess extensive information about their perceptions and concerns – an inherently time-consuming undertaking,

---

[12]   IRGC, supra note 4.

[13]   Id. at 10.

[14]   For nanotechnology, for example, see Gary E. Marchant et al., "Big Issues for Small Stuff: Nanotechnology Regulation and Risk Management," 52 JURIMETRICS 243–77 (2012).

and one for which new procedures may need to be adopted. Risk management involves identifying and evaluating a range of alternatives before selecting appropriate risk managers and management measures, and frequently entails actions to further international harmonization. Communication requires that authorities provide opportunities for meaningful stakeholder and public input. And all of these activities require exercises of judgment, not simple ministerial responses.

Even more important, the IRGC framework assumes that innovation will not outpace the risk governance process. Notions such as early warning, scientific and concern assessments, characterization and evaluation – indeed, even the choice of risk management measures – make sense only if the technology under consideration retains the same contours and remains at a similar level of development while those operations are being performed, or at worst evolves temporally in parallel to the risk governance process. If a technology changes form or diversifies into multiple forms, products and applications more rapidly than the process can be carried out, risk governance may never catch up.

Yet the IRGC assumption of parallel pacing is not satisfied on two counts: current technologies are "emerging" at an exponential pace, as noted above, while risk governance is slowed and hampered by systemic weaknesses in regulatory systems, as highlighted throughout this volume.[15] Many United States regulatory agencies – especially those with environmental responsibilities – are operating under outdated legislation, dating back as far as the 1970s. Such vintage statutes are often ill-suited to the issues posed by current technologies; for example, few provide for the kinds of societal input modern risk governance requires. In the current political climate, however, the chances that such legislation can be updated, or that new *sui generis* statutes can be adopted, are virtually nil. Exercises of regulatory authority also face strong opposition based on small-government and free-market ideologies and fear of the "nanny state."[16] In addition, many regulatory agencies and other oversight bodies face declining financial resources – rather than the increased resources that well-structured risk governance on the IRGC model would require – both because of efforts to constrain regulation and because of severe budget pressures.

Many scholars, including several of the editors and contributors to this volume, have called for "governance" or "new governance" approaches to oversight, to supplement or substitute for formal governmental

---

[15] For example, Marchant and Wallach, supra note 8.
[16] Bowman, supra note 5.

regulation.[17] Governance approaches rely on decentralizing regulatory authority among public, private and public-private actors and institutions; they also rely on softer forms of oversight, such as information disclosures, codes of conduct and certification mechanisms. Such arrangements can be adopted and revised more rapidly than formal regulations; they are also more easily adopted in anticipatory or experimental forms. Yet while governance approaches may help address the pacing problem, they have significant limitations: soft governance measures lack the enforceability of mandatory regulations; soft and private measures require voluntary participation, which may not be forthcoming (for example, the EPA's recent Nanoscale Materials Stewardship Program took several years to develop but produced very limited participation); and private codes of conduct may lack the legitimacy of public regulation, producing a lesser degree of public confidence and trust.[18]

1.2.2   Quality of Risk Governance

Apart from the sheer speed of their development and diffusion, the nature of current emerging technologies makes high-quality risk governance difficult to achieve. These technologies are protean: each technological family or platform includes numerous forms. For example, "'nanotechnology' is not simply one discipline, or family of techniques, but rather a vast range of disciplines, including engineering, materials science, biotechnology, medicine, physics, chemistry and information technology."[19] Each technology also encompasses numerous products or applications, and potentially many more.[20]

---

[17]   See, for example, Gary E. Marchant, Kenneth W. Abbott, Douglas J. Sylvester and Lyn M. Gulley, "Transnational New Governance and the International Coordination of Nanotechnology Oversight," in THE NANO-TECHNOLOGY CHALLENGE: CREATING LEGAL INSTITUTIONS FOR UNCERTAIN CHALLENGES (David Dana, ed., 2012); Kenneth W. Abbott, Gary E. Marchant and Elizabeth A. Corley, "Soft Law Oversight Mechanisms for Nanotechnology," 52 JURIMETRICS 279–312 (2012); Gregory N. Mandel, "Regulating Emerging Technologies," 1 LAW, INNOVATION, AND TECH. 75 (2009); LeRoy Paddock, "An Integrated Approach to Nanotechnology Governance," 28 UCLA J. ENVTL. L. AND POL'Y 251, 255 (2010).

[18]   See Abbott et al., supra note 17.

[19]   Graeme A. Hodge, Diana M. Bowman and Andrew D. Maynard, "Introduction: The Regulatory Challenges for Nanotechnologies," in INTERNATIONAL HANDBOOK ON REGULATING NANOTECHNOLOGIES (Graeme A. Hodge, Diana M. Bowman and Andrew D. Maynard, eds, 2010).

[20]   For examples from nanotechnology and synthetic biology, see Paddock and Masterton, supra note 3.

As a result, risk governance must contend with pervasive uncertainty as to the future course of technological development, undermining the logical progression of the risk governance process. In addition, the breadth of these platforms and the range of their impacts – as well as the overlaps and interactions among platforms – make emerging technologies poor fits with existing statutory allocations of regulatory authority: these features greatly complicate the crucial phase of risk governance in which oversight or risk management responsibility is assigned to the most appropriate regulatory scheme and agency – or more likely divided among multiple schemes and agencies, or perhaps assigned to a newly created scheme or agency. In most regulatory systems, moreover, no meta-regulator, other than the legislature, has the authority to make such determinations.

Uncertainty also characterizes the risks and benefits of current emerging technologies.[21] As Mandel notes, "One of the greatest challenges facing emerging technology governance is scientific uncertainty concerning the potential human health and environmental impacts of a technology."[22] For some technologies, such as synthetic biology, the implications are "not only difficult to predict but are fundamentally unknowable."[23]

Broadly conceived, the risks posed by emerging technologies extend well beyond traditional health, safety and environment issues. They engage "different ideas and values about the role of technology in society, assurances of safety, and equal sharing of risks and benefits,"[24] and other elements of diverse risk cultures. They also encompass broader impacts on social, economic and personal life, sometimes raising difficult ethical issues. Examples include the possibility of significant life extension through genetic and nanoscale therapies; radical enhancement of human capacities through a range of technological interventions; creation of new life forms; intrusions on personal privacy through nano and

---

[21] As Paddock and Masterton note, supra note 3, such uncertainty characterizes almost all new technologies; again, however, current emerging technologies appear unusually complex and uncertain.

[22] Gregory N. Mandel, "Emerging Technology Governance" (this volume, Chapter 3).

[23] Paddock and Masterton, supra note 3 (quoting Joy Y. Zhang, Claire Marris and Nikolas Rose, "The Transnational Governance of Synthetic Biology: Scientific Uncertainty, Cross-Borderness and the 'Art' of Governance," 3 (BIOS, The London School of Economics and Political Science, Working Paper No. 4, 2011), available at http://royalsociety.org/uploadedFiles/Royal_Society_Content/policy/publications/2011/4294977685.pdf (accessed July 7, 2013).

[24] Kuzma, supra note 11.

biosensors; and unjust economic and social outcomes due to differential access to valuable innovations.

Technologies characterized by this degree of uncertainty are particularly likely to produce deficiencies and failures in governance, what the IRGC calls "risk governance deficits."[25] Deficits are especially likely in the early phases of the risk governance process, in which information is gathered and risks are assessed and appraised. IRGC observes that risk governance failures during these phases are frequently caused by lack of sufficient factual information or misinterpretation of available information; failing to consult or to properly assess the concerns, risk perceptions and risk attitudes of stakeholders and the public; and failing to consider "Black Swans" – potential factual or social impacts outside normal expectations. Diverse, uncertain technologies may also lead to deficits in risk management, such as failing to adopt appropriate strategies, anticipate the side effects of management measures, address long-term risks and manage conflicts.

In addition, IRGC argues that when new developments occur within complex systems, significant risks and risk governance deficits are more likely to arise; for example, unpredictable risks may emerge from chance variations, and non-linear effects may arise when unanticipated thresholds are crossed.[26] Emerging technologies are evolving within multiple complex systems: the "scientific-industrial complex" of technology promotion, research and development, commercialization and global markets; the highly politicized "regulatory system" of national and subnational agencies, international institutions, and non-state actors and organizations exercising regulatory authority; the "attitudinal system" of public and stakeholder values, attitudes, social relationships and norms; and the social-ecological and earth systems in which health and environmental impacts unfold.

To address this array of governance problems, oversight of emerging technologies must be flexible and adaptable in the face of uncertain

---

25   IRGC, "Policy Brief, Risk Governance Deficits: Analysis, Illustration and Recommendations" (2010), available at http://www.irgc.org/risk-governance/irgc-risk-governance-deficits (accessed July 7, 2013).

26   IRGC, "The Emergence of Risks: Contributing Factors," 5–6 (2010), available at http://www.irgc.org/risk-governance/emerging-risk/irgc-concept-of-contributing-factors-to-risk-emergence/ (accessed July 7, 2013). Complex systems may, however, be more adaptable and resilient.

technological futures, risks and benefits;[27] resilient in the face of changing conditions and attitudes and possible surprises;[28] and robust in the face of threatening developments that have not been understood or anticipated.[29] These demands are not always consistent: for example, the decentralization of authority, multiple oversight bodies and soft measures of "new governance" may produce greater flexibility and resilience (especially through redundancy), but reduce robustness. More fundamentally, demands for adaptability, resilience and robustness place great stress on regulatory systems; as with the pacing problem, most systems fall far short.

1.2.3 Stakeholder Engagement

Numerous actors and organizations perceive themselves as having stakes in the benefits, risks and impacts of emerging technologies. These include scientific and technological researchers, business firms and industry associations, government officials, and civil society groups focused on issues from health, worker safety and environmental protection to privacy and economic justice. Private actors and organizations that engage in technology oversight through new governance arrangements have particular interests in the management of technological risks.[30] Societal groups with varying attitudes toward technology and risk also have stakes; in a real sense, so too do members of the general public, as consumers, as victims of potential risks or in other capacities. These actors increasingly demand information, transparency and voice,[31] and may challenge decisions in which they have not participated, making stakeholder engagement a necessity.

More fundamentally, as the IRGC risk governance framework suggests, neither technologies nor risks are exclusively objective, factual phenomena; they are in significant part socially constructed, based on subjective attitudes, understandings, values and cultures. Technologies

---

[27] Mandel, supra note 22; Marchant and Wallach, supra note 8; Leon Fuerth, "Operationalizing Anticipatory Governance," 2 PRISM 4, 31 (2011), available at http://www.ndu.edu/press/prism2-4.html (accessed July 7, 2013).

[28] IRGC, supra note 25, at 15; IRGC, supra note 4, at 17.

[29] Id. at 17. The IRGC recommends robustness-focused strategies where risks are "complex," with the underlying causal relationships poorly understood; it recommends resilience-focused strategies where risks are "uncertain," with technical data lacking or unclear.

[30] For examples from nanotechnology, see Abbott et al., supra note 17.

[31] Kuzma, supra note 11.

and risks for which attitudes and understandings vary widely are referred to as "ambiguous." For example, certain individuals and groups may understand a technological application such as genetically modified crops as a radical innovation with unknowable and potentially dangerous consequences; others may perceive it as an extension of familiar processes, such as plant breeding.[32] Similarly, certain individuals and groups may perceive particular potential impacts as threatening or ethically repellent; others may find them neither threatening nor ethically troubling. Risk governance that does not identify and take into account the relevant attitudes and concerns will not succeed.

Finally, modern regulatory policy, including risk regulation policy, views public communication, input and participation as essential. To cite just a few examples, the 2012 OECD recommendations on regulatory policy – approved by the governments of all developed nations – call for "open government," including transparency and communication, stakeholder engagement throughout the regulatory process, and open and balanced public consultations.[33] Executive Order 13563, issued in 2011, requires a regulatory process that "involves public participation," including "the open exchange of information and perspectives among … experts in relevant disciplines, affected stakeholders in the private sector, and the public as a whole."[34] The Principles for Regulation and Oversight of Emerging Technologies, issued by the White House Emerging Technologies Interagency Policy Coordination Committee on March 11, 2011, calls for "ample opportunities for stakeholder involvement and public participation;" these procedures are "important for promoting accountability, for decisions, for increasing trust, and for ensuring that officials have access to widely dispersed information."[35] And IRGC calls for "inclusive governance," based on an appreciation of the contributions stakeholders can make to risk governance; it recommends engaging wider circles of stakeholders as the technical, social and ethical issues raised by particular technologies become more extensive.[36]

---

[32]   IRGC, supra note 4, at 8–10 (noting differences in the subjective "framing" of genetically modified crops and nanotechnology).

[33]   OECD, Recommendation of the Council on Regulatory Policy and Governance (Annex, section 2), March 22, 2012, available at http://www.oecd.org/governance/regulatory-policy/ (accessed July 7, 2013).

[34]   E.O. 13563, Improving Regulation and Regulatory Review, January 18, 2011, 76 (14) Fed. Reg. 3821 (January 21, 2011).

[35]   http://www.whitehouse.gov/blog/2011/03/16/emerging-technologies-committee-lays-out-principles-guidance (accessed July 7, 2013).

[36]   IRGC, supra note 4, at 18.

Many regulatory statutes, schemes and agencies, however, lack strong provisions and procedures for stakeholder engagement. As Jennifer Kuzma notes, stakeholders and publics are normally limited to formal comment procedures once regulations have been drafted, and to *ex post* litigation if regulations are unsatisfactory.[37] This is a far cry from the ongoing engagement central to the IRGC process. Resource constraints also bind tightly here; active stakeholder and public engagement can be costly. At the same time, expansive stakeholder participation can delay the risk regulation process, and may even interfere with desired applications of expertise and judgment by risk managers. Where actors, organizations and societal groups view ambiguous technologies and risks with different understandings, concerns and values, moreover, consensus on risk governance measures becomes far more difficult to achieve. As in most areas of regulation, oversight bodies may be forced to make trade-offs between desired features.[38]

1.2.4 Coordination

In a complex regulatory system, the quality and even the pacing of oversight can suffer when a multiplicity of actors exercise regulatory authority without sufficient coordination. With no meta-regulator to manage the system as a whole, substantive issues or aspects of the risk governance process can fall through the cracks: each oversight body focuses on a narrow set of issues, viewing novel risks as someone else's responsibility.[39] In addition, oversight regimes can overlap. In these cases, at best, duplication of efforts saps scarce resources from the governance system and creates additional costs for the targets of oversight. Duplication can also lead to competition among oversight bodies, consuming additional resources, distracting organizations from their assigned missions and leading to conflict that complicates efforts at coordination. At worst, duplication can produce conflicting standards, creating severe problems for agencies, targets and stakeholders.

Strong risk governance processes like the IRGC framework require a high degree of coordination. Responsibility for each phase of the process must be clearly and appropriately assigned. For example, IRGC strongly recommends that risk assessment and risk management responsibilities

---

[37] Kuzma, supra note 11.

[38] On trade-offs as a characteristic problem of regulation, see Robert Baldwin, Martin Cave and Martin Lodge, UNDERSTANDING REGULATION 29, 32–5 (2d ed. 2012).

[39] IRGC, supra note 25, at 16.

be assigned to different bodies; both bodies should cooperate in other phases of the process, such as pre-assessment and communication.[40] IRGC also emphasizes that institutional responsibilities should be clearly assigned and accepted.[41] In its view, oversight bodies must take the broader institutional context into account:[42] gaps and overlaps resulting from dispersed and uncoordinated responsibilities are serious governance deficits that can undercut decision-making.[43]

Three types of coordination problem may plague governance of emerging technologies:

- First, governance of different forms, applications or impacts of a technology may fall to multiple governmental agencies. This is an almost inevitable product of the breadth of the current emerging technologies, as noted above. While bodies such as the Emerging Technologies Interagency Policy Coordination Committee can bring some order to such arrangements,[44] inter-agency coordination is a perennial problem.

- Second, governance of a technology may be multi-level, in the usual sense of involving multiple levels of government.[45] This is a particular problem within the European Union, where EU and national regulations and management bodies – and perhaps regional, local and international norms and agencies as well – often have overlapping jurisdictions and responsibilities.[46] It is less a problem in the United States, but oversight actions by state and local governments and any relevant international standards or international harmonization procedures add complexity to the oversight system.

- Third, a different multi-level governance problem arises when private and public-private bodies engage in oversight as part of new governance approaches. In the case of nanotechnology, for

---

[40]   IRGC, supra note 4, at 8.
[41]   Id. at 13.
[42]   Id. at 20–1.
[43]   IRGC, supra note 25, at 16.
[44]   For discussions of inter-agency coordination mechanisms, see Mandel, supra note 22; Kuzma, supra note 11; Joshua W. Abbott, "Network Security Agreements: Communications Technology Governance by Other Means" (this volume, Chapter 11).
[45]   Liesbet Hooghe and Gary Marks, "Unraveling the Central State, But How? Types of Multi-Level Governance," 97 AMER. POL.SCI. REVIEW 233–43 (2003).
[46]   IRGC, supra note 4, at 21.

example, current and recent programs that engage in soft or non-legally-binding forms of oversight have been sponsored by international organizations (for example, OECD working party on health and safety implications of manufactured nanoparticles); national agencies (for example, NIOSH recommended exposure levels for workers); multi-stakeholder public-private partnerships (for example, Responsible NanoCode); and private organizations dominated by business (NanoSafety Consortium for Carbon), researchers and civil society (Foresight Institute Guidelines), and business-civil society collaborations (Environmental Defense-duPont Nano Risk Framework).[47] Communication and other relationships between public regulators and non-state organizations are often weak, and traditional inter-governmental mechanisms are inapplicable.

In all these settings, the need for inter-institutional coordination is strong, but is often not satisfied.

1.3 OVERVIEW OF VOLUME

The editors asked the contributing authors to consider these and other problems relating to the oversight of emerging technologies, and to propose ways to respond to those problems. The result is a rich set of empirical examples, analyses and recommendations. This section provides a brief overview of the contributions of succeeding chapters.

Part I considers general issues of risk regulation and technology oversight. Braden Allenby begins by analyzing the complexity of technological systems – as illustrated by the railroad and its co-evolving clusters of technologies and social/political institutions – and the dynamics of technological change. Allenby argues that today's technologies – especially the "Five Horsemen" of nanotechnology, biotechnology, robotics, information technology and applied cognitive science – are not only revolutionary in scope, scale and speed of innovation; they constitute a fundamental shift in the state of the earth system.

Gregory N. Mandel focuses on the delicate problem of managing risks while still ensuring beneficial innovation, in the fields of biotechnology, nanotechnology and synthetic biology. He argues that one-size-fits-all command-and-control regulation is an ineffective response to this problem, and recommends a more flexible "governance" approach that

---

[47]   Abbott et al., *supra* note 17.

maintains sufficient governmental oversight to generate public confidence. Mandel's suggested strategy uses the uncertainty created by an emerging technology as an incentive to bring stakeholders together, early in the life of the technology, to agree on and implement a regulatory approach; the approach chosen will necessarily evolve over time in an iterated process.

LeRoy Paddock and Molly Masterton build on earlier work recommending an integrated governance system for nanotechnology, extending that approach to synthetic biology. They call for an adaptive system of government regulation that encourages experimentation and learning and continually adjusts risk management measures based on lessons learned. In addition, Paddock and Masterton recommend a multi-faceted governance approach that simultaneously deploys information disclosure mechanisms, codes of conduct, civil liability and public dialogue.

Marc A. Saner argues that governance should extend beyond methods for controlling emerging technologies, especially where those technologies are changing faster than regulatory agencies can foresee or adapt, and where they raise difficult ethical issues. In such cases, governance should also incorporate social adaptation mechanisms, as is done in the area of climate change. Saner argues that a focus on adaptation enables holistic thinking, facilitates experimentation, helps ensure maximization of benefits, and allows for valuable action if decisions on oversight are deadlocked. In addition, a focus on adaptation may improve control approaches by accepting their limits.

Timothy F. Malloy focuses on government policies that promote new technologies, which often fail to take into account health, safety and environmental risks, social impacts and other unintended consequences, as illustrated by the case of the fuel additive MBTE under the Clean Air Act. Malloy analyzes a range of governance deficits that lead to unanticipated consequences. In response, he calls for government to integrate precaution with promotion. More specifically, he recommends the incorporation of technology assessment procedures that consider societal implications at the legislative stage of technology promotion.

Gary E. Marchant and Wendell Wallach emphasize the coordination problem in technology governance, especially where multiple public, private and public-private bodies engage in soft governance initiatives. Governance schemes like these are typically created in piecemeal fashion, and often operate independently. To address the problem that there is normally no meta-regulator to induce cooperation among diverse initiatives, Marchant and Wallach call for the establishment of "issue managers" for each technology of concern. An issue manager would collect information, promote coordination, provide a forum for stakeholders and manage communication. After consideration of possible models, the authors propose a "governance coordinating committee" structure that would actively engage governmental and non-governmental stakeholders.

Part II of the volume considers country- and technology-specific governance issues and developments. Diana M. Bowman presents an Australian perspective. The governments of Australia and the state of Victoria, in particular, have aggressively regulated or mandated a range of technologies; these range from seat belts and motorcycle helmets to *in vitro* fertilization, embryo research and genetically modified foods. Bowman concentrates on the limits of government intervention, highlighting the resistance to activities that raise the specter of a "nanny state," recent moves to limit intervention through regulatory reviews and the "better regulation" agenda, and Australia's less aggressive response to nanotechnology. Her chapter contains a rich array of examples from the Australian regulatory experience.

Jennifer Kuzma uses the "tortoise and hare" metaphor to review the history of regulatory efforts to keep pace with the rapidly evolving technology of genetically modified crops. She finds that oversight largely succeeded in keeping pace temporally with technological developments, through a series of oversight mechanisms. However, many stakeholders find the resulting oversight system insufficient to ensure health and environmental protection in the long term, as well as insufficiently transparent and participatory. Kuzma thus argues for "proper pacing," in which oversight both keeps pace and remains responsive to potential harms and public demands for information and voice.

Rachel A. Lindor and Gary E. Marchant explore innovative regulatory programs being implemented by US agencies for molecular diagnostics. These products represent a technology transition from a simple, inexpensive medical assay to a much more complex, expensive test that now plays a central role in directing medical interventions and prediction. The regulatory framework established for the old model of diagnostic tests does not fit the sophisticated new technologies. With no statutory changes imminent due to the inertia of the legislative process, two US agencies are experimenting with innovative new approaches for the regulation and reimbursement of molecular diagnostics within the pre-existing legal framework. This example demonstrates the potential for creative innovation within existing legal frameworks.

Joshua W. Abbott examines a traditional regulated industry – telecommunications – that is also marked by rapid technological change. He focuses on an oversight procedure used in the review of license applications for foreign-owned communications networks to deal with potential national security, law enforcement and similar risks. This procedure is designed primarily to deal with the coordination problems raised by such broad policy concerns. The Federal Communications Commission defers to the Executive Branch, which exercises its oversight responsibility through ad hoc interagency teams. In addition, applying a new governance technique, interagency teams sometimes negotiate voluntary network security agreements with applicants; terms of these agreements become conditions of FCC licenses.

Preben H. Lindøe, Michael Baram and John Paterson examine risk regulation of offshore oilrigs and platforms. They first compare the regulatory approaches of the United Kingdom and Norway, which moved to performance-based rules following a series of accidents in the 1970s and 1980s, with that of the United States, which continued to impose technically detailed prescriptive regulations. They then examine the ways in which all three countries are rethinking their approaches in the light of the 2010 BP Deepwater Horizon Prospect disaster. Their analysis focuses on factors such as political and legal culture, social values and industrial-labor relations.

Gary E. Marchant draws overall conclusions regarding innovative governance models for emerging technologies in the concluding chapter of the volume. He concludes that there are no "golden bullets" that provide a perfect solution to the governance challenges of emerging technologies, but that the types of innovative governance described in this volume are helping us to muddle through.

PART I

General approaches

2. The dynamics of emerging technology systems

Braden Allenby

2.1 INTRODUCTION

Asking whether current governance models and institutions are adequate to address the dynamics of emerging technology systems raises several underlying questions. In particular, if previous governance mechanisms were adequate, and one wants to argue that current ones may not be, one needs to demonstrate that either governance mechanisms have changed, or that there is something unique about today's emerging technology systems that have, in some way, reduced the effectiveness of the applicable governance mechanisms. Law, regulation, and institutional structures – governance mechanisms, in some loose sense – indeed change all the time, but it is difficult to argue that they have, as they evolve, perversely become less rather than more effective in themselves.

Indeed, the classic institutions of the Enlightenment – the state, urban systems, law and regulatory structures, international trade, human rights law, the laws of armed conflict, and much else – not only continue to dominate global governance patterns, but appear to continue to adapt to new developments relatively gracefully. Thus, for example, the laws governing conflict and war, which include the Hague Conventions, the Geneva Conventions, the Chemical Weapons Convention, the Biological Weapons Convention, relevant international humanitarian law, and much else, continue to evolve as new technologies, such as cyber weapons, drones, or directed energy weapons, are developed.[1] Arguendo, it must therefore be the case that there is something about today's emerging technology systems – which for purposes of this chapter can be taken to

---

[1] See generally Brian Orend, THE MORALITY OF WAR (2006); Jeff A. Bovarnick et al., LAW OF WAR DESKBOOK 165–9 (2011); Braden R. Allenby, "The Implications of Emerging Technologies for Just War Theory," 27 PUBLIC AFFAIRS QUART. 49 (2013).

include at least nanotechnology, biotechnology, information and communication technology (ICT), robotics, and applied cognitive science – which differs in significant ways from past experience.

From a historical perspective, this is not a trivial claim. Many technology systems have had major impacts in their historical era; indeed, it is virtually the hallmark of a technology system that it destabilizes the economic, social, cultural, and institutional environment within which it develops.[2] Moreover, it is generally the case that each generation overemphasizes the degree of change that it experiences, in part because of the immediacy of the stresses and changes to which it is exposed, and in part because, history being settled, it is easy to underestimate how difficult and unpredictable life was at the time. Nonetheless, there are several developments that might support an argument that emerging technologies today are different not just in degree, but in kind, from those that humans have experienced in the past:

1.   The planet that humans live on today is different from anything else we are aware of because it is the first terraformed planet – that is, the first world we know of anywhere that has been shaped, indeed transformed, by the deliberate activities of a single species.[3]

2.   The scope and scale of technological change is unprecedented. Previous waves of technological change have generally involved a few core technologies – steam and coal, for example, or railroads and automotive technology. But today, technological evolution is occurring across the entire technological frontier: core emerging technologies, such as nanotechnology and ICT, are not just important in themselves, but are enabling technologies whose effects extend across all technology systems, as well as society and culture itself (the latter dynamic especially true of information technologies).[4]

3.   The levels of social, cultural, and technological complexity, especially when blended with the global impacts of human systems, are

---

  [2]   See generally Nathan Rosenberg and L.E. Birdzell, Jr., HOW THE WEST GREW RICH: THE ECONOMIC TRANSFORMATION OF THE INDUSTRIAL WORLD (1986); Chris Freeman and Francisco Louçã, AS TIME GOES BY: FROM THE INDUSTRIAL REVOLUTIONS TO THE INFORMATION REVOLUTION (2001).

  [3]   Editorial, "Welcome to the Anthropocene," 424 NATURE 709, 709 (2003); see generally Braden R. Allenby, RECONSTRUCTING EARTH (2005).

  [4]   Braden R. Allenby, THE THEORY AND PRACTICE OF SUSTAINABLE ENGINEERING 251 et seq. (2012).

unprecedented, and, among other things, complex adaptive systems require very different institutional and psychological approaches than current practice supports.[5]

4. The human itself has become a design space. While humans have always changed themselves in many ways – from consuming intoxicants of all kinds, to medical practices through the ages, to educational systems both informal and formal – the direct interventions that are now possible, combined with accelerating advances in fields such as neuroscience, make virtually all aspects of the human, including emotional and ethical intelligence, potentially subject to design.[6] Physically, it is not unusual to hear medical researchers confidently predicting that individuals born today may well have a lifespan of 150 and beyond, with a high quality of life (others, of course, disagree).[7]

5. The rate of change of technology is accelerating.[8] Many people are familiar with Moore's Law, which holds true for ICT in generalized form: functional capability of ICT products doubles roughly every 18 months. They are less aware that the same holds true for biotechnology – Carlson Curves, for example, show that the cost of sequencing bases is decreasing, and biosynthesis productivity, in terms of base production per person per day, is increasing, in the same geometric manner as in the case of ICT.[9]

It is not just that each of these factors might suggest a change from past patterns. Rather, it is that they are all operating together, in ways that synergistically increase the impact, speed, and depth of change, that support a perspective that the current era is indeed, as the historian J.R. McNeill entitled his 2000 book, *Something New Under the Sun*.

An additional complication is that these processes are nonlinear. So, for example, there has always been technological change, but for virtually all of human history until the last several hundred years, it did not occur at such a rate as to disturb the basic state of being human, which was essentially subsistence and a relatively short lifetime. As Clark summarizes it, "The average person of the world of 1800 was no

---

[5] See generally Joseph A. Tainter, THE COLLAPSE OF COMPLEX SOCIETIES (1988); Braden R. Allenby and Daniel Sarewitz, THE TECHNO-HUMAN CONDITION (2011).

[6] See generally Allenby and Sarewitz, supra note 5.

[7] See generally Aubrey De Grey and Michael Rae, ENDING AGING (2007).

[8] See generally Ray Kurzweil, THE SINGULARITY IS NEAR (2005).

[9] Allenby, supra note 4, at 208.

better off than the average person of 100,000 BC ... [l]ife expectancy was no higher in 1800 than for hunter-gatherers: thirty to thirty-five years."[10] But with the Industrial Revolution came what economic historians call "The Great Divergence": income per person and technological capability exploded in industrial countries, while undeveloped countries remained at the historical level of subsistence.[11]

Waves of technological evolution there may have been, but the rate of technological change by and large remained similar enough to the rates of social and institutional change so that decoupling was not a major problem. But especially since World War II, and the push that the subsequent Cold War gave to technological innovation (because technological competence, especially in military and security domains, was a major arena of superpower competition), the rate of change of technology has accelerated. And, in part because technology is autocatalyzing, it continues not just to advance, but to accelerate its advance. The result is that, arguably, the rate of technological change has now decoupled from the legal, social, and institutional mechanisms that have in the past been an important part of assuring social and cultural stability. Thus, the critical dynamic that this book attempts to address is whether, if this is the case, more agile and rapid "soft law" alternatives can be developed that to some extent enable more effective adaptation than the formal, and more conservative and inertial, tools of traditional hard law and regulation. But before the specifics of these issues can be addressed, it is necessary to understand something of the dynamics of technology systems.

2.2 TECHNOLOGY AS CHANGE IN EARTH SYSTEM STATES: RAILROADS

To note that the Earth is increasingly a terraformed planet has become almost passé. Illustrative examples abound. The Earth's emission spectrum, for example, is no longer determined by the natural characteristics of the planet, such as reflections from clouds, emitted infrared radiation, planetary heat content and composition, and atmospheric physics and chemistry; it also includes electric lights and heat island effects from

---

[10]   Gregory Clark, A FAREWELL TO ALMS: A BRIEF ECONOMIC HISTORY OF THE WORLD 1 (2007).
[11]   See generally Rosenberg and Birdzell, supra note 2; Allenby, supra note 4, at 12.

urban centers, television and radio broadcasts, and the like. Anthropogenic global climate change offers another example; virtually by definition, this would not be a concern if it weren't that humans had so affected a natural system. The bioavailable nitrogen cycle at virtually all scales is dominated by the results of human technology (especially fertilizer production and use), as is the phosphorous cycle. Biology is an interesting example: while most conservation biologists argue that the world is experiencing a "crisis in biodiversity" as human activity causes extinction levels to accelerate, bioengineers are busy constructing new forms of life, leading some to claim that there is no overall loss of biodiversity, but merely a shift from "evolved biodiversity" to "engineered biodiversity." All of these examples, and many others that will suggest themselves upon a moment's reflection, strongly suggest that the current era is not, as is frequently claimed, beset with a series of "problems" to be "solved" as much as a fundamental shift in earth system state, from a condition where humans are only one species among many, to one where a single species has grown to dominate the planet. It follows that the search for "solutions" is misplaced, and that the human species will be engaged in a dialog with systems such as the climate, atmospheric chemistry and physics, and the carbon, nitrogen, phosphorous, hydrologic, and other natural cycles, so long as populations and economic activity remain at anywhere near current levels. Impacts on these complex and inter-related system can be redistributed, but growing human influence cannot be eliminated.[12]

Key to these impacts, of course, are the technologies that humans have developed, especially over the several hundred years since the Industrial Revolution and its concomitant demographic, agricultural, cultural, social, and institutional changes. Technologies in this broad sense do not simply represent physical artifacts; rather, they are integrated with, coupled to, and co-evolve with virtually all human, social, and cultural systems. These integrated systems thus have a far different, and higher, degree of complexity than is usually realized.[13] A historical example may clarify this somewhat abstract point. Consider the railroad, by now a familiar, perhaps even dated, technological component of the global landscape. But for those in Western Europe or the United States who experienced it in the 1840s and 1850s as it began its rapid expansion phase, it was the harbinger of profound change in virtually all domains of

---

[12] See generally Allenby, supra note 4.
[13] See generally Wiebe E. Bijker, Thomas P. Hughes and Trevor Pinch, THE SOCIAL CONSTRUCTION OF TECHNOLOGICAL SYSTEMS (1997).

life. Take, for example, the concept of time that for most people is so fundamental it is never even noticed, much less questioned. Prior to railroads, local times varied unpredictably: London time was four minutes ahead of Reading, for example, and 14 minutes ahead of Bridgewater; in the US as late as 1850 there were more than 200 local times.[14] But the time fragmentation that was entirely serviceable when connections between towns was by horse, walking, or canal – so that slippage of a few minutes was entirely unnoticeable given the time lags involved in transportation and communication – becomes entirely unacceptable when running a railroad, characterized by regional scale integrated networks requiring a uniform, and precise, system of time. (Interestingly enough, exactly the same dynamic is at work in the chips in computers and mobile phones; the time network in that case is measured in nanoseconds, reflecting the needs of the network integrated across the chip.)

Moreover, this railroad time example nicely illustrates the interplay between cultural, industrial, and technological networks: the adaptation to uniform systems of time was neither smooth nor immediate. In the United States, for example, an intermediate stage consisted of uniform times maintained by individual train companies, rather than integrated across society: at one point, the train station in Buffalo, served by three railroad companies, had three different clocks; Pittsburgh, a larger station, had six. Part of the issue – which seems strange to moderns completely acculturated to today's time system – was that no one knew what a regional, national, and global system of integrated time would look like. So, for example, at one point US rail companies used 80 different times, coordinated but rather confused.[15] Indeed, it was not until 1883 that rail firms in the US standardized their times to four time zones, and 1918 that regional standard time gained legal recognition in the US.[16] In short, the system of "industrial time" that now governs across the world reflects the needs, characteristics, and requirements of railroad technology.[17] But a network technology such as a railroad system does not just require coextensive time technology; it also requires a communication technology that is fast enough, and geographically disperse enough, to manage such a network – in other words, it requires

---

[14] Wolfgang   Schivelbusch,   THE   RAILWAY   JOURNEY: THE INDUSTRIALIZATION OF TIME AND SPACE IN THE 19TH CENTURY 41–4 (1997); Alan Beattie, FALSE ECONOMY: A SURPRISING ECONOMIC HISTORY OF THE WORLD 203 (2009).

[15] Beattie, supra note 14, at 203.

[16] Schivelbusch, supra note 14, at 44.

[17] Rosenberg and Birdzell, supra note 2, at 149.

something like the telegraph, which in point of face was frequently laid along the same rights of way as the railroad technology it enabled.[18] As early as 1849, the New York and Erie Railroad was pioneering the use of telegraph technology; so great was the need for such a communication overlay that only five years later most companies had adopted it.[19]

Railroads had profound effects on economic and power structures. Earlier forms of transport were slow and expensive, meaning that only lightweight, high value, non-perishable items were worth shipping any distance. But railroads created regional and national economies of scale, and made trade in commodities, even perishable agricultural commodities, possible (in 1851, the Northern Railroad of New York ran the first cold car; by 1857 the first shipment of beef was made from Chicago to the East Coast). Prior to the railroad, economies were essentially local, especially with bulk or perishable items; after the railroad, economies were national and global – Midwest wheat was shipped to Chicago and then to New York by rail, then on to Europe by steamer.[20] The era of monopolies and trusts – Big Oil, Big Sugar, Big Tobacco – rode across continents on rails of steel (Big Steel, of course). Prices for commodities converged, first in national markets, then internationally.[21] This represented a huge political and economic shift as economic power passed to industrial firms from agriculture; more subtly but perhaps more fundamentally, so did cultural authority.[22]

These fundamental economic, political, and cultural shifts were mirrored in the landscape. Continental interiors, such as the American Midwest, had been settled by Europeans, but the model was generally subsistence farming at low intensity. After being reached by rail technology, however, market access was dramatically enhanced both physically and economically, and new markets were created as perishable agricultural products could be shipped to the East Coast markets. Industrial agriculture became economically feasible. Accordingly, railroads transformed the landscapes they touched; Chicago grew, and in turn structured the American Midwest physically and environmentally,

---

[18]   Arnulf Grubler, TECHNOLOGY AND GLOBAL CHANGE 62 (1998).
[19]   Beattie, supra note 14, at 203.
[20]   William Cronon, NATURE'S METROPOLIS: CHICAGO AND THE GREAT WEST 109 et seq. (1991).
[21]   Beattie, supra note 14, at 200 et seq.
[22]   See generally Leo Marx, THE MACHINE IN THE GARDEN: TECHNOLOGY AND THE PASTORAL IDEAL IN AMERICA (1964); David E. Nye, AMERICAN TECHNOLOGICAL SUBLIME (1994).

because of railroads.[23] (This is not simply of historical interest; consider what the economic effects of a meaningful railroad network across Africa might mean in terms of development – and, also, in terms of social, economic and cultural disruption.)

Railroad technology had deep psychological implications as well. Railroad technology was seen by many contemporaries as a discontinuous change from previous forms of transport, not just extending, but obliterating the sense of place and rhythm that "more natural" and far slower transportation technologies such as the horse and carriage and canal had encouraged. The sense of natural distance was distorted as the words of Heinrich Heine, written in 1843 with the opening of new rail lines across France, illustrate:

> What changes must now occur, in our way of looking at things, in our notions! Even the elementary concepts of time and space have begun to vacillate. Space is killed by the railways, and we are left with time alone. ... Now you can travel to Orleans in four and a half hours, and it takes no longer to get to Rouen. Just imagine what will happen when the lines to Belgium and Germany are completed and connected up with their railways! I feel as if the mountains and forests of all countries were advancing on Paris. Even now, I can smell the German linden trees; the North Sea's breakers are rolling against my door.[24]

There was a widespread belief that traveling at the fantastic speed of 25 miles per hour would kill the passengers, and that railroad technology was against the obvious will of God: "If God had designed that His intelligent creatures should travel at the frightful speed of 15 miles an hour by steam, He would have foretold it through His holy prophets. It is a device of Satan to lead immortal souls down to Hell."[25] In some ways, this reflects a strong tendency to characterize change, especially technological change, as blasphemic – consider, for example, Prince Charles' complaint in an article in the *Daily Telegraph* in 1998 that genetically engineering agricultural crops was evil because it was intervening in "realms that belong to God and God alone."[26] More prosaically, passengers on the new railroads complained of being treated like baggage rather

---

[23] See generally Cronon, supra note 20.

[24] Quoted in Schivelbusch, supra note 14, at 37.

[25] Ohio School Board (1828) quoted in Nye, supra note 22, at 57.

[26] Jeff Randall, "Prince Charles Warns GM Crops Risk Causing The Biggest-Ever Environmental Disaster," DAILY TELEGRAPH, August 12, 2008, available at

than as individuals,[27] which, given the complaints about air travel that pepper the Internet these days, is another response that does not appear to have changed over the years. A more important psychological effect, perhaps, is reflected in the change in cultural perspective that the railroad encouraged, especially in the US, as the ideal individual shifted from being a farmer or frontiersman, to the technological; from Jeffersonian agrarianism, to a technology-driven New Jerusalem.[28]

Railroads also significantly impacted geopolitics, especially in large countries such as the US, although the effects were more subtle than sometimes realized. Most people, for example, are aware of the immediate military advantage that a strong railroad infrastructure provided. This became clear during the American Civil War, where the technology provided the US not just the ability to transport men and war materials to areas of immediate need more effectively and promptly, but also was a critical infrastructure supporting greater productivity and industrial efficiency. As always, however, it was not just the technology system, but the technology system as embedded in the larger society and economy: the railroad density (track per unit land area) of the North was three times that of the South, but so was the canal density, indicating something of the systemic advantages of the US, and, more generally, an industrial versus an agrarian economy.[29]

A similar if less known example of the geopolitical importance of railroad technology is provided by the rise of Prussia. In 1815 when the Congress of Vienna concluded the Napoleonic Wars, Prussia was only a minor state among the Great Powers of the time, particularly France, Russia, and the Austrian Empire. But all of Europe experienced a reaction to the conservative absolutism that the Congress attempted to impose, and popular revolts broke out across the continent in 1848. Prussia had been investing heavily in rail technology during the 1830s and 1840s, with the result that, in Prussia at least, the uprisings were controlled in part because the Prussian military used railroads to rush troops from trouble spot to trouble spot, quickly shifting military strength

---

http://www.telegraph.co.uk/earth/earthnews/3349308/Prince-Charles-warns-GM-crops-risk-causing-the-biggest-ever-environmental-disaster.html (accessed July 7, 2013).

[27] Schivelbusch, supra note 14, at 121.

[28] See generally Marx, supra note 22; Nye, supra note 22.

[29] John Keegan, A HISTORY OF WARFARE 305 (1993); Daron Acemoglu and James Robinson, WHY NATIONS FAIL 351 (2012).

to where it was most needed.[30] For Prussian leaders such as Helmuth von Moltke, this confirmed the strategic importance of railroad technology. Financial innovations such as the Prussian Railway Fund enabled construction of militarily critical lines that were not commercially viable, essentially enabling dual use of Prussian railroad networks. In another example of the dual use approach, Prussian commercial railroad cars were designed so that in addition to their routine commercial purpose they could carry soldiers, horses, military supplies, and military equipment if necessary. Nowhere in Europe was rail technology so coupled to military planning: Prussian mobilization plans, for example, were based on railroad infrastructures (which, in turn, had been designed to support mobilization). For example, upon mobilization regiments were assigned to a specific railroad which had already been designed to enable rapid and efficient loading of military trains.[31] It would be a mistake to think that the approaches taken by the US North in the Civil War and by Prussia – the one a creation of private firms supported by significant state incentives, the other much more a product of direct state planning – were either self-evident or easy. In contrast to Prussia, for example, the French, like the American South, were lukewarm on railroads; the rate of expansion of rail in France during the 1840s was half that of Prussia.[32] An even greater contrast was presented by Russia and Austria, where the ruling elites deliberately stifled the spread of railroads because of the possibility that their still predominant feudalistic structures would be destabilized.[33] Attempts to halt powerful technology systems are seldom if ever successful in the long run, however. In the case of Europe, this became apparent in 1866 at the Battle of Koniggratz, when Prussia, heretofore a minor state, stunned the Austrian Empire. Railroad technology was not, of course, the only factor, but it was a major one: the Prussians managed to transport 197,000 men and 55,000 horses to the front with unprecedented rapidity using railroads, catching the Austrians by surprise in part because of their lack of familiarity with railroad technology.[34] Combined with other technologies such as the needle gun, arguably the most advanced rifle in Europe, superior leadership and administration, and highly advanced training, Koniggratz marked the rise

---

[30]   Max Boot, WAR MADE NEW 125 (2007); Alleny and Sarewitz, supra note 5.

[31]   Boot, supra note 30, at 124 et seq.

[32]   Geoffrey Parker, THE CAMBRIDGE HISTORY OF WARFARE 239 (2005).

[33]   Daren Acemoglu and James A. Robinson, WHY NATIONS FAIL 226 et seq. (2012).

[34]   Boot, supra note 30, at 142.

of a new European power, Prussia, and the beginning of the end of a reigning European empire, Austria.

2.3 KONDRATIEV WAVES AND TECHNOLOGY CLUSTERS

The railroad example is useful because it is on the one hand such a mundane and familiar technology, and on the other hand, it plays such an important co-evolutionary role in so many different domains. But it is only one example: co-evolved networks of social, economic, cultural, theological, institutional, moral and political patterns associated with a core technology have characterized a number of foundational technology systems. Indeed, economic historians have developed the concept of "long waves" of innovation (sometimes called "Kondratiev waves"). Such waves consist of a core technology or set of related technologies, such as steam and coal, which create, and co-evolve with, fundamental cultural, social, institutional and economic changes, just as in the case of railroad technology. Just as railroads required substantial change in time and communication technologies, each core technology supports "technology clusters." (Obviously, what the boundaries are between core technology and technology clusters, and which particular technology is defined as "core," is somewhat arbitrary; moreover, virtually all the "core technologies" are actually networks of innovations, artifacts, and institutional change – "railroad technology," for example, consists of many, many different technologies integrated into complex artifacts and physical networks).

Using this approach, economic historians have used economic data to identify a number of Kondratiev Waves. The first Kondratiev Wave was the beginning of the Industrial Revolution, around 1750 or so in the United Kingdom until around 1840 to 1850, and was defined by textile technology. The associated technology cluster, however, included numerous advances in mining, agriculture, transportation, and manufacturing, and, from an institutional perspective, the rise of factory capitalism and thus a whole set of new institutional, financial, and managerial structures.[35] The associated changes, from rapid urbanization to the growth of a new class, the urban proletariat, were just as profound as those accompanying railroad technology, extending across virtually all domains

---

[35] See generally David S. Landes, THE UNBOUND PROMETHEUS (2nd ed., 2003).

of culture and even religion – remember William Blake's lines in "Jerusalem":

And did the Countenance Divine,
Shine forth upon our clouded hills?
And was Jerusalem builded here,
Among these dark Satanic Mills?

Bring me my Bow of burning gold;
Bring me my Arrows of desire:
Bring me my Spear: O clouds unfold!
Bring me my Chariot of fire!

I will not cease from Mental Fight,
Nor shall my Sword sleep in my hand:
Till we have built Jerusalem,
In Englands green and pleasant Land.

What is generally considered the second Kondratiev Wave, based on railroad and steam technology, lasted from about 1840 to 1890 or so, and was followed by a third, based on steel, heavy engineering and electricity, lasting from about 1890 to 1930. The fourth Kondratiev Wave, with core technologies including the automobile, petroleum, and aircraft, is generally understood to have lasted from about 1930 to 1990. Finally, many experts, such as Freeman and Louca,[36] postulate a fifth Kondratiev Wave based on a core of information technologies, including computerization of the economy, beginning about 1990 and extending through to the present (I will argue below that the fifth Kondratiev Wave is actually far more complex and fundamental than just information technology, important as that is). Obviously, any such binning exercise is somewhat fraught and artificial, but the general idea of clusters of technology, carrying with them institutional, organizational, economic, cultural and political changes, is a useful one.[37]

Moreover, although the economic literature does not emphasize this, each of these waves does not just destabilize and restructure human systems at all scales. They do the same for natural systems. Railroads open up continental interiors to industrialized agriculture, which, supported by advances in industrial chemistry and production of artificial fertilizers, creates the potential for dramatic increases in global human

---

[36]   See generally Chris Freeman and Francisco Louca, AS TIME GOES BY: FROM THE INDUSTRIAL REVOLUTIONS TO THE INFORMATION REVOLUTION (2001).

[37]   See generally id.; Landes, supra note 35.

population. Internal combustion engine technology combined with a psychologically potent automotive technology fed by petroleum technology re-engineers the atmosphere. Urbanization enabled by the shift away from agriculture and towards manufacturing and then service industries changes a dispersed, rural species into an urban species, with concomitant changes in land use, energy use, material flows, and cultural perspectives on such concepts as "nature" and "wilderness." The "natural sublime" of the early Enlightenment becomes the "technological sublime" of the modern.[38] In short, each technology cluster is not just a change in technology, or in the structure of capitalism; it is a change in Earth systems. Without conscious intent, each technology cluster is, in fact, a step towards a terraformed planet – and not just any terraformed planet, but one that bears the imprint of the technologies, and the choices, and the designs, upon which humans settle.

It must be emphasized that, although this approach views these clusters of integrated change in terms of technology, what is involved here is not "technological determinism" – the idea that technology dominates and determines all other domains. Rather, these are co-evolving systems, and while it is helpful to adopt a particular focus – say, technological systems – to try to understand them, it is always important to remember that no single perspective is more than partial. Thus, for example, the Marxist emphasis on economic class and means of production is a very useful way to frame many analyses, but economic determinism – the idea that economics determines the course of nations and of history – is too simplistic and offers only a partial explanation of the way events unfold in the real, highly complex, world. Similarly, while environmental considerations are often important in decision making, the efforts by environmentalists and sustainability advocates to frame all human activity in terms of environmental considerations – environmental determinism – also fails. Among other things, environmental determinism is blind to the complexity of human societies (often called "wicked problems," a term of art[39]).

It cannot thus ever be forgotten that every wave carries with it fundamental change across all domains. Since technology is the lens, this is perhaps most obvious in the way new technology systems require new management and financial structures. Thus, specialized professional

---

[38] See generally Nye, supra note 22; Meyer H. Abrams, NATURAL SUPER-NATURALISM: TRADITION AND REVOLUTION IN ROMANTIC LITERATURE (1971).

[39] See Horst W.J. Rittel and Melvin M. Web, "Dilemmas in a General Theory of Planning," 4 POLICY SCIENCES 155, 161–7 (1973).

managerial systems and associated "Taylorism" industrial efficiency techniques characterized the heavy industry cluster, while a far more networked, flexible structure began to evolve during the information cluster.[40] Automobiles and the advent of mass consumption of large and expensive items such as white goods (heavy consumer durables such as refrigerators, washers and driers, and air conditioners) required a new way of financing consumer purchases, since very few potential consumers were able to accumulate all the cash required for the entire purchase price of such expensive products. Thus, means to provide widespread consumer credit – including not just new financial mechanisms, but new institutions to operate those mechanisms – had to be developed.[41] Widespread consumer credit, in turn, generates an ability to consume that was beyond imagining for generations of earlier human beings, and, among other things, in turn raises new and challenging environmental considerations.

These examples also illustrate several general principles of technological evolution. Most importantly, any technology of enough significance to be interesting will inevitably destabilize existing institutions, power relationships, social structures, reigning economic and technological systems, and cultural assumptions. This inevitably creates opposition to new technologies, both by conservative social forces (remember how Austrian and Russian elites stifled railroad technology), and by threatened economic interests (such as the Recording Industry Association of America suing thousands of people in an effort to defend a technologically obsolete business model). To the extent such opposition is successful, historical experience suggests that it will not halt the evolution of technology; examples of failure include the Chinese and Japanese attempts to limit gunpowder technology, which, although successful in the short term, left both cultures open to subjugation by Western naval forces with gunpowder technology. Similarly, many forces in Europe have been very aggressive in attempting to ban genetic engineering in agriculture, but outside Europe it has been one of the most rapidly adopted technologies in agricultural history. Especially given today's globalized culture, technological leadership is likely to simply pass to other cultures where opposition is less effective, especially if the technology has significant military, strategic or economic potential.

---

[40]   See generally Manuel Castells, THE RISE OF THE NETWORK SOCIETY (2nd ed., 2000).

[41]   Freeman and Louca, supra note 36, at 238.

A second important point is that projecting the effects of technology systems before they are actually adopted and become embedded in their social and cultural context is not just hard but, given the complexity and reflexivity of the systems involved, essentially impossible.[42] The time structure that eventually resulted from railroad technology is a good example. Not only was it not the time structure of pre-railroad American agrarian society or European cultures, but it was also not predictable: globally ordered time frames reflecting a fragmentation across the world of a 24-hour system were not just unpredictable *a priori*, they were difficult to conceive in social and cultural systems that had neither any need, nor any concept, of unified and ordered temporal frameworks at a planetary scale. It is useful to hypothesize about potential effects of technology, but always important to remember that one is dealing with scenarios, rather than predictions. Among other things, this implies that great care should be taken if current policy is based on hypotheticals about technology futures. This is especially true given historical experience: virtually no powerful technology has not been attacked when introduced, often by existing economic, ideological and political interests masquerading as representatives of a broader public interest.

Successful efforts to block technological evolution would be of less concern if it were not for the historical observation that cultures that develop technology, and create conditions that facilitate technological evolution such as developing frameworks that enable technology to react upon itself and so accelerate its own evolution, tend to gain ascendancy over competitors.[43] Cultures that attempt to block technology even for reasons that appear desirable will, all things equal, eventually be dominated by those that embrace it. This obviously poses an unhappy dilemma: if a culture wishes to maintain dominance, must it develop (or, alternatively, both develop and deploy) all technologies where it is capable of so doing? If this is the case, does it imply that ethical judgments about technologies move over time to the lowest common denominator? There are no good answers to these questions, but they do indicate the likelihood that in a highly competitive global environment, where many cultures are jostling for position, technological evolution will be difficult, if not impossible, to stop. The more subtle question of how technology systems can be moderated in the age of global elites becomes an important research question under such circumstances (even assuming

---

[42]   Rittel and Webber, supra note 39, at 167.
[43]   See generally Laurence D. Harrison and Samuel Huntington, CULTURE MATTERS: HOW VALUES SHAPE HUMAN PROGRESS (2000).

that sufficient knowledge and institutional capability exists to intervene in such complex systems in ways that produce desired outcomes with a high probability of success).[44]

2.4 THE NEXT KONDRATIEV WAVE: THE FIVE HORSEMEN

As noted above, the literature has suggested that the core technology of the current Kondratiev Wave is information and communications technology, or ICT. Unfortunately, however, this appears to understate the challenge considerably. It is not just that economic history suggests that technology systems are far more complex and potent across virtually all human and natural domains than usually contemplated by analyses and studies in various disciplines. Rather, it is that the history of technological evolution is an entirely inadequate warning given the wave bearing down on us, for rather than just one or two enabling technologies undergoing rapid evolution, we have the Five Horsemen: nanotechnology, biotechnology, robotics, ICT, and applied cognitive science.[45] It is, in other words, not just a matter of ICT; it is a matter of fundamental and unpredictable evolution across the entire frontier of technology.

Taken together, these technologies are in some ways the logical end of a long chapter of human effort to gain mastery over natural, built, and human systems, one that is thousands of years old. Nanotechnology extends human will and design to the atomic level, while biotechnology and synthetic biology extend it across the biosphere.[46] ICT provides the ability to create virtual and synthetic realities at will, to migrate functionality to information rather than physical structures, and, increasingly, to build smart environments that integrate the physical and information worlds as desired. Robotics is not only a major domain where the integration of the physical and the informational occurs, but also a means by which human functionality can be diffused across technology platforms. Applied cognitive science not only informs ICT and robotics, but enables vastly more powerful diffusion of cognition across integrated techno-human networks with unpredictable effects.[47]

---

[44]   Allenby and Sarewitz, supra note 5, at 159 et seq.
[45]   See generally Joel Garreau, RADICAL EVOLUTION (2004); Allenby, supra note 4.
[46]   John Robert McNeill, SOMETHING NEW UNDER THE SUN 193 (2000).
[47]   See generally Edwin Hutchins, COGNITION IN THE WILD (1995); Allenby and Sarewitz, supra note 5.

The question as to whether current conditions differ in any meaningful way from past experience must always be asked, but in some ways it is too superficial. Past experience can almost always inform current conditions, but does so best when both similarities and differences are acknowledged. In this case, for example, it is clear that technological change, even at a very fundamental level, is not new to human experience, but, to the contrary, is a fairly quotidian experience at least since the Industrial Revolution. But there also appear to be several factors which may be different about the current wave. First, the Five Horsemen suggest that the current scope and scale, and accelerating rate, of technological evolution are unprecedented. Second, the complexity of interactions has increased as the human impacts scale from the predominantly local, to the regional and global, accompanied by increased integration and coupling of natural, built, and human systems. Finally, as will be discussed below, the integrated impact of the Five Horsemen is not just to operate at the heretofore unprecedented level of a terraformed planet, but to make the human into a design space in ways that have not been possible before. Taken together, then, there is at least a reasonable argument that current conditions are, indeed, unprecedented and must be managed with knowledge of the past, but without assuming that historical experience is definitive under current conditions.

A brief review of the core technologies constituting the Five Horsemen is in order. Consider nanotechnology, which, in extending human will and design to the atomic level, constitutes in some ways the culmination of the human project to gain ascendency over physicality. Nanotechnology may appear to simply be a rebranding of chemistry, which, after all, has long dealt with phenomenon at the nanoscale (a nanometer is one billionth of a meter), and it is true that many devices, especially in electronics, have long had design elements at the nanometer scale. Most definitions of nanotechnology, however, include three components: (a) design and manipulation at a scale under 100 nanometers; (b) deliberate manipulation of materials or constructed devices at that scale; with, (c) design and functionality that depend on the new and novel properties or materials and artifacts at that scale. It is the emphasis on design and function at the nanoscale that differentiates nanotechnologies from chemistry.

Because many people conflate "nanotechnology" with "nanomaterials," it is worth noting that the two are not the same. Current nanotechnologies do indeed often focus on nanomaterials, but the domain includes growing development and deployment of simple but active nanostructures such as sensors and drug delivery device. Looking further, there is increasing interest in not just more complex nanostructures containing thousands of interacting components, but, especially at the border between nano-technology and biotechnology, integrated nanosystems with complexity at the scale of the human cell (including biological systems such as DNA where appropriate).

Like nanotechnology in the physical realm, biotechnology extends human domination across the biosphere at all scales; as is the case with other emerging technologies, technological evolution in biotechnology is accelerating dramatically. For example, DNA synthesis, and synthetic DNA production, capability in terms of molecule length are advancing exponentially, while the cost of gene synthesis and base sequencing are falling equally rapidly.[48] This is leading to dramatic decreases in the cost of human genome sequencing which, in turn, will have substantial impacts in areas as varied as human health, the practice of medicine, drug development, privacy, and insurance. Obviously, the implications are not limited to humans: such technologies enable genetic engineering of crops and food animals to increase yield, reduce pesticide consumption, reduce demand for land for agriculture and energy production, enable plant growth under saline and arid conditions, increase hardiness, and support a healthier human diet. Indeed, biotechnology at some point will enable the development of designer foods engineered to produce greater nutri-tion and health benefits, which can be produced in factories; the economic, cultural, social, and institutional implications of such a shift are obviously enormous, yet most people don't realize the initial relevant research is already under way.[49]

As with many emerging technologies, biotechnology merges cutting edge science and engineering; an example is the growth of the field of "synthetic biology," sometimes called "plug and play biology" (carried to a logical conclusion by tools such as "Biobricks," a "Registry of Standard Biological Parts"[50]). Among other things, synthetic biology envisions the creation of standard biological components, circuits and modules, with standard interfaces, so that they can be mixed and matched in target organisms to provide desired functions, thus enabling design extensions beyond existing biological systems in new and completely anthropogenic ways. Moreover, because synthetic biology involves design choices

---

48   Allenby, supra note 4, at 208.
49   See Carolyn S. Mattick and Braden R. Allenby, "Cultured Meat: The Systemic Implications of an Emerging Technology," in PROCEEDINGS OF THE IEEE ANNUAL SYMPOSIUM ON SUSTAINABLE SYSTEMS AND TECHNOLOGY (May 2012).
50   Biobricks, available at http://partsregistry.org/Help:An_Introduction_to_ BioBricks (accessed July 7, 2013).

towards explicit ends such as economic efficiency, it results in a very different biosphere than non-directed evolutionary processes would generate, but the implications of such shifts cannot be evaluated until it is actually perceived (and not simply demonized) by the relevant communities. More subtly, perhaps, biology shifts from being a "natural science" to being an engineering discipline driven by economics and social needs. That relevant institutions and disciplines are slow to recognize, much less adapt to, such trends is simply another demonstration of the way that accelerating technological evolution has decoupled from existing psychological, institutional, and disciplinary systems. It is not just law and regulation that are unable to keep up with technological change.

Robotics is a more applied technology domain than either nanotechnology or biotechnology. In fact, often robotic systems are integrations of other technology – one example is the experimental robots that use a neuronal mass from rats as part of their control and behavioral system.[51] Moreover, because they represent a more applied domain, much of robotics has expanded as a result of their economic role. Robots also have a longer pedigree than some emerging technology areas; older machine tools in some ways were robots, and so are newer, more versatile industrial robots. Recently, robotic development has responded to broader opportunities. In particular, some cultures facing rapidly approaching demographic challenges as the ratio of working age individuals to seniors continues to shift are looking to robotic systems, rather than immigration, to help support their productivity (Japan is the leading example). Moreover, because of approaching labor shortages and a growing elderly population, much of the recent work in robotics in Japan has focused on developing robots that can substitute for humans in elder care and nursing.[52] The other domain where growth in robotic systems has been obvious is in the military, where robots that can disarm bombs on the ground, and provide constant surveillance and attack capability from the air ("unmanned aerial vehicles" or UAVs) are proliferating.[53] For the United States, a major military robotic power, this is driven by two trends: the need to keep military casualties to a minimum to avoid

---

[51] Kevin Warwick, "Robots with biological brains," in ROBOT ETHICS: THE ETHICAL AND SOCIAL IMPLICATIONS OF ROBOTICS 317, 317–22 (Patrick Lin, Keith Abney and George A. Bekey, eds, 2012).

[52] See generally Patrick Lin, Keith Abney and George A. Bekey, ROBOT ETHICS: THE ETHICAL AND SOCIAL IMPLICATIONS OF ROBOTICS (2012).

[53] See generally Peter W. Singer, WIRED FOR WAR: THE ROBOTICS REVOLUTION AND CONFLICT IN THE 21ST CENTURY (2009).

losing domestic support for military activity, and the need to substitute capital (robots) for labor in production processes – less need for expensive workers in factories, fewer war fighters in harm's way in combat at a time when long term demographics indicate that there will be fewer young people available to recruit.[54]

Partially because it is so applied, and partially because it is in some ways more anthropomorphic than most technologies, robotics has generated a fair amount of literature regarding questions of ethics. While the results are mixed to this point,[55] two trends stand out. First, like law and regulation, the framing and analysis of ethical issues involving emerging technologies badly lags the cutting edge of the technology system itself. Second, much of the initial response to new technologies, ethical as well as legal, will represent defensive moves by existing institutions and ideologies, often using dystopian projections of potential effects of technology to protect existing economic interests and worldviews. This analysis is reinforced by looking at the impacts of robotic systems in military operations, where it is clear that the explicit laws of armed conflict, and international humanitarian law, are challenged by the emergence of new technologies, and the military and security strategies, institutions, and practices based on them.[56] For example, a US UAV drone may be controlled by military personnel, by intelligence personnel (for example, CIA), or by a private contractor: different rules of engagement, and legal regimes, apply in each case. Both of these domains, then, arguably provide case studies of the decoupling of technology systems from existing hard law systems, and argue for the development and deployment of more agile, and sophisticated, soft law instruments. Note that it does not necessarily mean that the hard law regime should be rejected completely; only that, under current circumstances, it is probable that hard law alone is inadequate to provide rational governance of emerging technology systems.

ICT is both an emerging and a critical infrastructure technology. At the level of function, the accelerating growth of ICT capability is captured by the famous relationship first suggested by Gordon E. Moore, one of Intel's founders, in 1965. Moore's Law observed that the number of integrated circuits per chip was doubling approximately every two years,

---

54   Allenby, supra note 1.
55   See generally Lin et al., supra note 52.
56   See generally Allenby, supra note 1; Carolyn S. Mattick, Braden R. Allenby and George Lucas, 2012 CHAUTAUQUA COUNCIL FINAL REPORT: IMPLICATIONS OF EMERGING MILITARY/SECURITY TECHNOLOGIES FOR THE LAWS OF WAR (in press, 2013).

and indeed this trend has continued to the present. More importantly, most other measures of ICT technology, from pixels per dollar to Internet traffic to information storage capacity per hard disk, have followed a similar doubling trend. Most non-technologists do not appreciate how powerful such a consistent doubling trend can be, especially when accompanied by comparable drops in cost per unit function: computing performance per unit cost has also been doubling approximately every two years.[57]

Such integrated rates of growth are, of course, critical because if growth lagged in one of the technology components – software, or chip speed, or Internet capacity – it would eventually limit the ability of ICT technology systems as a whole. When, on the other hand, all the critical technologies develop at roughly the same speed, it is possible to maintain systemic coherence in technological evolution. It is worth noting that this is achieved in the electronics industry not by chance, but by intent (within the constraints of antitrust and other relevant law): the industry relies heavily on "roadmaps" – explicit, agreed upon projections of technology development in particular domains such as information storage, chip performance, and the like, created in industry consortia and organizations – that work to assure alignment of goals, technological developments, and interfaces across these complicated technology systems. The implications for soft law – that technological complexity is manageable, but may require new institutional and cultural relationships (such as cooperating on roadmaps with firms that are also one's competitors in the marketplace) – are obvious, and worth considering. Most legal scholars, for example, are not familiar with roadmapping and other processes by which technologists manage their difficult realms.

Especially where a technology such as ICT constitutes critical infrastructure, such a rapid rate of change has emergent properties across many industries, and social and cultural practices. An obvious example is the way that traditional newspapers and their associated reporting and information gathering functions have been undercut by the Internet and blogs. Another example is the rapid replacement of landline telephone service with mobile phone services. In 2009 in the US, for example, customers were abandoning landlines at a rate of 700,000 per month, marking a significant change in communication technology. Whereas in 2005 only 5% of American households were mobile phone only, by 2008 over 20% were – in fact, in that year the number of mobile-only families

---

[57] Kurzeil, supra note 8, at 56 et seq.

exceeded landline-only families.[58] Such shifts have huge impacts: mobile phones, for example, enable social networking services that are difficult and far less convenient, if available at all, over landline systems. It is no coincidence that the explosive growth in social networking services coincided with this shift: Facebook was introduced in 2004, and passed its billionth account in late 2012; Twitter was introduced in 2006 and by 2010 was an important contributor to geopolitical upheavals (for example, the abortive Iranian revolution and the "Islamic Spring" in Tunisia, Egypt, and elsewhere).

Some implications are more subtle. For example, in the United States government regulation has for years taxed landline telephone service in order to subsidize rural landline telephone service. Accordingly, if shifts to mobile service reduce the available subsidy, rural areas might become even more depopulated than they are now given that communication services are often necessary for any sort of employment in today's economy. Alternatively, purely private mobile phone service might be extended to all but the most remote areas based on changed economics: while a rural landline is an expensive system to deploy and maintain, a few cell towers are far less expensive, and can provide equivalent or better service. It is not just that hard law is lagging technological evolution even when it is as evident as it is in this instance; it is that the technology is morphing even more rapidly than the fundamental structure can appreciate. A mobile phone, after all, is not really even a telephone any more: it is an information device with ever more functionality, and with ever more capability to, for example, reduce privacy and enable a security state. Obsolete subsidies are perhaps the least of the problem, yet even in that case, responses are delayed and confused.

Importantly, it is not just explicit communication that accelerating technological evolution in the ICT domain affects. Culture itself is an information process, and it changes as the enabling ICT technology does. At the most obvious level, this dynamic is apparent in the rise of types such as the networked "digital native" and the gamer – individuals who have grown up with powerful virtual reality and networking ICT systems, and have simply adopted them as part of the ambient environment. Indeed, the availability of powerful search engines and the integrated nature of the Internet mean that every individual now has essentially the accumulated information of human history at her or his fingertips, with effects that are likely profound, but difficult to determine at this early stage in the evolution of human cognition to reflect changes in ICT. More

---

[58]   Allenby, supra note 4, at 217.

subtly, the shift from a world where information was relatively scarce, to one where it is abundant and powerful, obviously will have substantial economic and cultural impact, but is as yet poorly understood. Indeed, most people are unaware of how significant and recent this shift is: the ex-CEO of Google has noted that every two days humans produce as much information as was created between the dawn of civilization and 2003. Whether this is technically accurate has, of course, been the source of much argument – what, for example, constitutes "information" as opposed to "junk noise"? Nonetheless, the basic point – that humanity has tipped from a history where information was scarce, into a presence where it is beyond abundant, which psychological, social, institutional and cultural implications that are not even perceived, much less understood – is clear.

Moreover, one does not need to be a practicing postmodernist to understand that ICT mediates reality. Thus, for example, powerful video and virtual reality technologies, combined with rapid urbanization, are subtly changing the way that people understand and relate to cultural constructs such as "nature" and the "environment." Superficially, this can be seen in the substitution of passive consumption of information for actually experiencing "nature" outdoors.[59] More than this, however, people begin to perceive "nature" as it appears in television and cable programs, with animals that are always active and scary, in locations that are always beautiful and sublime. These are, of course, constructed images, which in order to compete with other video and Internet fare, must be constantly made more enticing, active and engaging, and because people are more urban, they are less and less frequently compared to external realities. The result is a growing gap between the conceptualization of "nature" gained from ICT and what's actually out there. More subtly, "nature" shifts from being a reality outside the human, to being a human construct that is increasingly designed, and increasingly – and appropriately, given such a framing – a reflection of human preferences.

Another potent field of technology usually included in the Five Horsemen is applied cognitive science ("cogsci"). As with all the technology systems mentioned above, cogsci integrates both technological and scientific domains, and acts as both a technological domain in its own right, and an underlying competence for other technology systems (for example, many developments in robotics arise from

---

[59] See generally William Cronon, ed., UNCOMMON GROUND: RETHINKING THE HUMAN PLACE IN NATURE (1995).

advances in cogsci). Definitional issues are especially difficult in this area, however. How, for example, should one categorize fundamental advances in search engines, which in one way are "simply" a dramatic extension of human memory, but in another way appear to create a cognitive system that includes not just most of recorded human experience, but also seem to lay the groundwork, when combined with social networking systems and personal information appliances of rapidly expanding competence, for emergent cognition at unprecedented scales (for example, crowd sourcing)? Similarly, augmented cognition, or "augcog," is a rapidly expanding area of research and product development where important elements of human cognition are dispersed across technology systems, creating integrated techno-human networks from which cognitive behavior emerges.[60] This need not be mysterious; most automobile companies today are developing technologies for autonomous vehicles by building cognitive functionality into the car, rather than relying on the driver to provide it.[61] The potential implications of advances in augcog, and in cogsci generally, are profound, especially when viewed through the lens of military operations and national security: when and how, for example, will it be possible to hack such integrated techno-human systems and, beyond that, perhaps to hack the human brain directly?

Such observations indicate how, without denigrating the profound effects of previous Kondratief Waves, the current wave may result in more radical psychological, institutional, social, and cultural change. It is not just the Earth at the scale of regional and global systems, but the human itself, that is in the early stages of becoming a human design project. While humans have always had a strong tendency to link with their environment, including technologies as they were developed, the Five Horsemen dramatically accelerate this reflexive process. So, for example, some predict the achievement of "functional human immortality" within a few decades as a result of the continued doubling of technological capability especially in biotechnology and ICT.[62] While many experts view such predictions as unlikely, there is also a growing consensus that substantial extensions of the average lifespan, with a high quality of life, are achievable in the near future: the IEEE *Spectrum*, a mainstream technical journal, concluded in a series of articles in 2004 that using "engineered negligible senescence" to control aging would

---

[60]   See generally Hutchins, supra note 47; Allenby and Sarewitz, supra note 5.
[61]   "Look, No Hands," THE ECONOMIST (Tech. Quart.), September 1, 2012, at 17–19.
[62]   Kurzweil, supra note 8, at 198 et seq..

allow average ages of well over 100 within a few decades.[63] Even in this latter case, the demographic, political, economic, and social stresses that such a development would entail are obviously substantial. Others, however, are optimistic. Roco and Bainbridge, for example, assert in their 2003 NSF report entitled "Converging Technologies for Improving Human Performance:" "With proper attention to ethical issues and societal needs, converging technologies could achieve a tremendous improvement in human abilities, societal outcomes, the nations's productivity, and the quality of life."[64]

2.5 CONCLUSION

History indicates that changes in foundational technology systems are profoundly destabilizing to existing psychological, institutional, social, cultural, and economic frameworks, and especially so to institutions, such as formal law and regulation, that are designed to support social stability. Under conditions of rapid change, the strengths of such conservative institutions can rapidly become weaknesses, as the cycle time of institutional change lags behind, or even decouples, from the changes occurring in economic markets and technological systems. This creates a need to develop more agile and adaptive tools, to be used in conjunction with more traditional approaches, which enable rational and ethical management of change in highly uncertain and rapidly evolving environments. So-called "soft law" tools, ranging from codes of conduct to roadmaps to informal standard-setting, may offer one such possibility.

---

[63] "Engineering and Aging," 41 IEEE SPECTRUM 10, 31–35 (2004).
[64] Mihail C. Roco and William Sims Bainbridge, eds, CONVERGING TECHNOLOGIES FOR IMPROVING HUMAN PERFORMANCE ix (2003).

3. Emerging technology governance

Gregory N. Mandel[1]

3.1 INTRODUCTION

A wondrous range of emerging technologies is expected to transform society. Biotechnology, nanotechnology and synthetic biology, to name a few, are anticipated to revolutionize fields as diverse as health care, agriculture and energy. These innovations may produce wonder drugs that target and destroy cancerous tumors, energy generation that reduces greenhouse gas emissions and molecular machines that manufacture products cheaply, cleanly and without waste. On the more concerning side, these technologies also could produce toxic substances that cause cancer or new organisms that disrupt ecosystems.

The development and governance of emerging technologies are inevitably and dynamically intertwined: these technologies cannot develop without providing researchers freedom to innovate, but too much freedom can lead to a calamity that forecloses future opportunity. Insufficient protection could lead to excessive or unknown human health and environmental risks and undercut public confidence. But excessive regulation could limit the development of an extremely promising technology and foreclose potentially great social, health, environmental and economic benefits. The emerging technology governance challenge requires designing an innovation environment that simultaneously permits pursuit of a promising innovation's anticipated benefits while guarding against its potential risks. This challenge is made more difficult by the reality that the risks of a technology often cannot be fully understood until the technology further develops.

This chapter proposes a governance system for managing this delicate balance for emerging technologies. Historically regulation has evolved reactively around relatively mature industries. Most regulation has proven

[1] Portions of this chapter are based on Gregory N. Mandel, "Regulating Emerging Technologies," 1 LAW, INNOVATION & TECH. 75 (2009).

remarkably unyielding to evolution, even in the face of recognized limits and flaws. The new governance model proposed here seeks to both move the point of first governance earlier in a technology's development and to enable the governance structure to evolve after its formation. It envisions a proactive, flexible form of governance – more of a governance process rather than intractable regulatory rules.

One obstacle to this goal is that new technologies are often met with highly polarized debates over how to manage their development, use and regulation. Prominent examples include nuclear energy and genetically modified foods. Proponents of a given technology often argue for promoting rapid development of the technology, unfettered by what they view as unnecessary and costly regulation. Opponents will advocate a stringent regulatory regime to protect against the potential human health and environmental risks of the technology. This polarization usually results in a long period of fractured – and rarely productive – debate. Eventually after many societal and political resources have been squandered, some form of regulation is instituted. This regulation tends to become the permanent status quo, tenaciously resistant to change, even as scientific information and the technology at issue evolve.

Scientific uncertainty is one contributor to the polarization and gridlock that often surround emerging technology governance and the regulatory ambiguity that results. Instead of allowing scientific and regulatory uncertainty to produce stagnation, however, it may be possible to leverage such uncertainty to achieve a more socially beneficial outcome. Uncertainty can be distasteful to all parties involved in an emerging technology – it can create fear and concern among members of the public and public interest groups, challenges and criticism of regulatory agencies, and limitations on industry plans for investment and development. Rather than being a source of division, it may be possible to harness this common concern. The emergent stage in particular, with a high degree of uncertainty and a low degree of attachment to any status quo, can present a unique opportunity to bring together diverse stakeholders to produce a collaborative governance process rather than a resource-draining adversarial battle.

The first part of this chapter discusses the rationale and challenges for developing an emerging technology governance framework. The second part provides particular recommendations for a generalized emerging technology governance system.

3.2   EMERGING TECHNOLOGY DEVELOPMENT

Though the opportunities for an emerging technology may appear almost boundless, these opportunities cannot be achieved if a technology is not developed in a secure manner that maintains public confidence. To understand emerging technology development and governance, the analysis here draws examples from various technologies at different stages of technological and commercial maturity, with a particular focus on biotechnology, nanotechnology and synthetic biology. Examining this range of technological development provides insight into how to manage emerging technologies more generally so as to simultaneously leverage potential benefits while guarding against potential risks.

3.2.1   A Brief Introduction to Biotechnology, Nanotechnology, and Synthetic Biology

Biotechnology, and particularly agricultural biotechnology for the purposes of this discussion, involves the purposeful transfer of one or several genes from one species to another in order to provide enhanced traits. For example, agricultural crops can be genetically modified to include genes that make them pest-protected or herbicide-resistant. The biotechnology discipline developed in the 1970s. Field tests of agricultural biotechnology began in the 1980s, and genetically modified crops were first commercialized in the 1990s. Genetically modified plants currently represent a dominant source of soybean, cotton and corn production in many countries, particularly the United States. Livestock genetically modified to improve food traits, animal health or produce products for human therapeutic use are close to commercial production as well.

Nanotechnology involves a variety of activities designed to manipulate matter at the atomic scale. Building matter from the atom up allows for more precise and complex configuration of material and permits the production of materials with different physical, chemical and biological properties than previously possible. Certain materials, for example, become stronger or more flexible, change their conductivity or develop biological and antimicrobial properties at the nanoscale. Such developments have widespread potential application in health care, medicine, energy, environmental sciences, electronics, optics and other applied materials sciences. Nanoscale manipulation was first achieved in the 1990s, and the first commercial products entered the market in the 2000s. However, the major advances, as well as areas of potentially greatest risk, are yet to come.

Synthetic biology entails applying engineering techniques to biology to permit the purposeful design of new organisms piece by piece. Synthetic biology will permit genes and other DNA fragments to be integrated together like building blocks, producing a living entity with a desired combination of traits, much as one can assemble a car with different features by putting together different individual pieces. The design of an organism through synthetic biology can include both the redesign of existing natural biological systems to have enhanced or novel qualities and the original construction of new biological systems that never existed in nature. Unlike traditional biotechnology, synthetic biology will permit the purposeful assembly of an entire organism, not just the transfer of one or several genes. Synthetic biology is in the most nascent stage of the technologies discussed: databanks of gene fragments are being developed and several of the steps necessary to engineer new organisms have been achieved, but a fully synthetic organism has not yet been built. Synthetic biology could allow better disease detection and treatment, more efficient energy production, new means of environmental protection and bio-remediation, and improved computer technology.

These and other emerging technologies will reshape our society in ways that cannot be fully understood. The goal of emerging technology governance is figuring out how to harness the power of such technologies while adequately guarding against their risks.

3.2.2   Regulation versus Governance

Both the type and magnitude of the benefits and risks created by emerging technologies are uncertain. The myriad unknowns presented by rapidly developing technological advancements create a considerable challenge for governance. The widely varied contexts and characteristics of these technologies mean that traditional one-size-fits-all command-and-control regulation generally will not be either efficient or effective. Designing a more flexible governance system that can respond to changing knowledge and information is necessary to optimally handle the benefits and risks of emerging technologies.

Governance, as opposed to regulation, does not mean that technological risks are not taken seriously and managed carefully. The failure to handle technological risks appropriately can have significant deleterious effects on human health and the environment and can limit future technological development. Addressing potential risks early and transparently is critical to the long-term success of emerging technologies.

The early stages of genetically modified food development in the United States provide a poster-child example of how significant public concern over a technology – and the perception that it is not being managed properly – can thwart technological development. In 1998, Monsanto CEO Robert Shapiro told his shareholders that the commercial introduction of genetically modified seeds was "[t]he most successful launch of any technology ever, including the plow."[2] Just one year later, a Deutsche Bank report titled "GMOs Are Dead" stated: "The term GMO has become a liability ... GMOs, once perceived as the driver of the bull case for this sector, will now be perceived as a pariah."[3] Despite the United States' current position as a leader in genetically modified food development, public reaction against GMOs has forced several large biotechnology companies to shelve some of their plans for further developments, such as research into genetically modified wheat. Public concern about genetically modified food products in many European countries has been even more virulent and has limited their introduction there as well.

Nanotechnology may face similar challenges as concern about appropriate regulation to protect human health, safety and the environment in light of potential nanotechnology risks is growing. A number of interest groups and commentators have called for substantial regulatory revisions in order to manage nanotechnology's risks; some have gone so far as to call for a nanotechnology moratorium. Similar pleas were and continue to be made concerning biotechnology. Public concern over nanotechnology (like biotechnology before it) has not been mollified by governmental agencies, whose response to potential nanotechnology risks has been perceived as slow and limited, displaying a degree of complacency not justified by current scientific understanding (again, like biotechnology before it).

Though ignoring technological risk can clearly be detrimental, identifying precisely how to respond is also usually not clear. There are often strong calls for enacting entirely new regulatory regimes or substantially overhauling existing laws to respond to new technological risks. Many such proposals for regulatory reform in light of emerging technologies, however, are neither particularly realistic nor useful. Overhauling health, safety and environmental laws, for example, is a remarkably expensive task and one that generally lacks sufficient political support. Even serious consideration of substantial legislative changes would involve a costly,

---

[2]   Paul Brown and John Vidal, "GM Investors Told to Sell their Shares," THE GUARDIAN, August 25, 1999.

[3]   Id.

resource-draining, lengthy and highly uncertain process with no guarantee of an outcome that is more protective or efficient than the existing structure.

On the opposite extreme from those who argue for regulatory redesign are others who advocate a free market, or fully voluntary, approach to emerging technology governance. One common variant on this theme is the refrain that regulatory change is unnecessary because emerging technologies and their attendant risks are no different from previous concerns. Such claims were raised in the context of biotechnology, based on the reasoning that farmers have selectively bred plants for modification for generations, and in the context of nanotechnology, based on the rationale that non-engineered nanoparticles have been around since the beginning of time. Though careful consideration must be paid to whether emerging technologies actually present new risks, some concern is legitimate in many cases. Modern biotechnology, for example, presents new risks because it enables a much broader array of genetic traits with much greater taxonomic divergence to be incorporated into a new organism than was ever possible through conventional breeding.[4] Similarly, intentionally engineered nanoparticles have new electrical, chemical and biological properties that can cause them to have different exposure and risk profiles and to exist in the environment for longer than non-engineered nanoparticles.[5]

Though market forces and non-governmental action can provide valuable controls in certain regards, they also cannot fully substitute for mandatory requirements. As with regulatory overhaul, there is generally little public support for voluntary or self-regulatory approaches to emerging technologies. An appropriate degree of government oversight is particularly necessary to maintain public confidence in emerging technologies as many people are often largely unaware of them. For example, surveys reveal that more than 80 percent of the public have heard nothing or only a little about nanotechnology and synthetic biology.[6] If people

---

[4] National Research Council, ENVIRONMENTAL EFFECTS OF TRANSGENIC PLANTS 48–52 (2002).

[5] INSTITUTE OF MEDICINE, IMPLICATIONS OF NANOTECHNOLOGY FOR ENVIRONMENTAL HEALTH RESEARCH 32 (Lynn Goldman and Christine Coussens, eds, 2005).

[6] Project On Emerging Nanotechnologies, PUBLIC AWARENESS OF NANOTECHNOLOGY: WHAT DO AMERICANS KNOW? WHO DO THEY TRUST? (2008), available at www.nanotechproject.org/events/archive/public_awareness_nanotechnology_what_do/ (accessed July 8, 2013); Gregory N. Mandel, Donald Braman, and Dan M. Kahan, CULTURAL COGNITION AND

learn that technology risk management is substantially voluntary at the same time as they first learn about a technology, public concern would be expected to increase rapidly, as it did with biotechnology. The combination of low public awareness and polarizing debates makes the landscape for socially appropriate development of nascent technologies even more challenging.

In contrast with advocates for regulatory overhaul, moratorium, free market and status quo approaches, the proposal presented here recommends a different approach to managing emerging technologies. This approach seeks to turn some of the greatest challenges of these technologies – scientific uncertainty and regulatory disruption – on their head to create incentives for diverse stakeholders to work together on a new governance system. This strategy is based on recognizing that scientific and regulatory uncertainty creates problems for most interested parties involved with an emerging technology. Uncertainty creates fear and concern among the public, regulatory challenges for and criticism of regulatory agencies, and produces a problematic environment for technological investment and industry development.

It is critical for industry that the public not lose faith in a technology or its risk-governance system at early stages of technological development. Nanotechnology currently faces this challenge as early nanotechnology products have entered the market and a larger number of more complex next generation advances are soon to be commercialized. Synthetic biology will enter this stage over the next few years as scientists develop the means to build novel organisms, such as microorganisms designed for bioremediation or engineered to produce biofuels and industrial chemicals. Concern about technological risk and uncertainty about how a technology will be governed can make investors unwilling to invest in the technology as well as make it more difficult for firms to know how to proceed with research, development and commercialization.

Mutual concerns about uncertainty not only provide normally opposed stakeholders incentives to work together, but also can be exploited to produce agreement on a particular governance system. For instance, both those who believe that a particular emerging technology is relatively

---

SYNTHETIC BIOLOGY RISK PERCEPTIONS: A PRELIMINARY ANALYSIS (2008), available at http://papers.ssrn.com/sol3/papers.cfm?abstract_id=1264804 (accessed July 8, 2013); Dan Kahan, Paul Slovic, Donald Braman, John Gastil & Geoffrey L. Cohen, AFFECT, VALUES, AND NANOTECHNOLOGY RISK PERCEPTIONS: AN EXPERIMENTAL INVESTIGATION (2007), available at http://papers.ssrn.com/sol3/papers.cfm?abstract_id=968652 (accessed July 8, 2013).

risk-free and those who are extremely concerned about its risk may be able to agree on a governance framework that will respond to new scientific information as it develops. Those who believe the technology is relatively safe may be satisfied because they believe that no significant risks will be uncovered, and technological development will continue relatively unabated. Those who are more concerned may agree to such a framework because they believe that the technology's risks will be revealed, and the framework will respond with predefined protective measures. Obviously this brief account elides many details and challenges, but it provides a potential conceptual structure for working toward practical emerging technology governance. Efforts in this direction are likely to be more rewarding than another resource-draining adversarial battle over a new command-and-control regulatory regime.

Similarly buttressing such efforts, uncertainty in scientific knowledge and regulatory management at emergent stages often means that interests and organizations have not yet strongly vested around a particular system or status quo. The combination of common concern about uncertainty and lack of a status quo can create a window of opportunity for a broadly developed and widely supported new governance model. Due to its wider base of support, such a model could be instituted far earlier in a technology's development, providing assurance to the public and stability for industry. This window, however, does not remain open indefinitely. As regulatory decisions and investments in particular products and uses are made or relied upon, many stakeholders become less flexible.

New technologies place stress on existing regulatory systems – systems that are customarily designed to handle technology as it exists when the regulatory system was developed. Emerging technologies disrupt these systems. It is not surprising that advances as transformative as biotechnology, nanotechnology and synthetic biology raise substantial problems for existing, mature regulatory systems. These disruptions, however, can provide opportunities to illuminate problems with the current structure and to rethink how emergent technologies are governed.

3.3 EMERGING TECHNOLOGY GOVERNANCE

This section develops a series of emerging technology management practices that are consistent with the proactive and flexible governance approach introduced above. The new governance recommendations presented here are directed at developing a reliable, efficient, adaptive, transparent and participatory management system. The recommendations are aimed at achieving a number of goals that would likely receive widespread support for any technology governance system: protecting human health, safety and the environment; not unduly hindering the development of a nascent, promising technology; advancing scientific understanding of the technology and its risks; achieving governance that is adaptable as the technology and scientific understanding advance; allowing for widespread participation in governance; and maintaining public confidence in the emerging technology and its management system. The proposal seeks these goals through regulatory agency, industry and public interest group cooperation and management incentives rather than relying primarily on conventional command-and-control regulation.

Because of the variation and uncertainty in emerging technology development, there are inherent limitations to how precise an *ex ante* governance framework can be. These limitations, however, do not prevent the identification of significant parts of a general governance structure, with specifications and details that can be worked out for particular technologies and specifications as a technology develops and its properties and risks become better understood. A general management structure can provide a type of best practices for emerging technology governance. Such a structure would deliver needed assurance and protection for the public and greater certainty for industry, as well as save time and resources for the government.

The new governance proposals developed below include a variety of recommendations focused on six areas: (1) improving data gathering and sharing in the face of limited resources; (2) filling newly exposed or created regulatory gaps; (3) incentivizing strong corporate stewardship beyond regulatory requirements; (4) enhancing agency expertise and coordination; (5) providing regulatory adaptability and flexibility; and (6) producing broad, diverse stakeholder involvement. The following sections detail these governance recommendations. This discussion is not intended to be comprehensive but to sketch out how the new governance proposals could produce a system that is more protective of human health and the environment, more efficient for industry and taxpayers, and promotes responsible technology development.

3.3.1 Data Gathering

One of the greatest challenges facing emerging technology governance is scientific uncertainty concerning the potential human health and environmental impact of a technology. For example, the US Environmental Protection Agency (EPA) has identified the need for greater scientific information on nanotechnology concerning "chemical identification and

characterization, environmental fate, environmental detection and analysis, potential releases and human exposures, human health effects and
ecological effects."[7] A primary focus of governance should be on
gathering all available data, developing as much new useful information
as possible and providing incentives for data reporting and development.
Greater data will provide the government, scientists and the public with a
better understanding of the applications and risks associated with a
technology.

At a basic level, the public, citizen groups and industry should place
pressure on public sources to increase funding for studies on the human
and environmental exposure and risks posed by emerging technology
products. Often the government substantially underfunds exposure and
risk research in relation to the funding available for technology development. Many commentators agree that this has been the case for nanotechnology, although funding for research into nanotechnology hazards
has recently increased. Public agencies should also take a lead role in
identifying research needs.

Regulatory agencies should take advantage of any existing authority to
encourage or require the development and production of further scientific
information. Agencies can also develop voluntary consultation programs
even where they lack authority to mandate reporting. Such policies can
provide firms with strong incentives to comply or risk consumer backlash. For example, although consultation with the Food and Drug
Administration (FDA) on the commercialization of most genetically
modified food products is voluntary, the FDA believes that it has been
consulted prior to the introduction of all new genetically modified food
products.

More innovative approaches to improve data gathering are also available, such as a model proposed for nanotechnology.[8] Rather than the
largely one-size-fits-all current regulatory scheme, nanotechnology substances and uses could be classified on a scale as being of negligible,
low, medium or high concern. The classification would be based on a
variety of relevant risk-related characteristics, such as a substance's size,
structure, coating, solubility and ease of transport in the body, as well as
its toxicity and expected exposure to humans or the environment. For
any given nanotechnology product or use, the extent of both pre- and

---

[7] Environmental Protection Agency, CONCEPT PAPER FOR THE
NANOSCALE MATERIALS STEWARDSHIP PROGRAM UNDER TSCA
20–21 (2007).
[8] See Gregory N. Mandel, "Nanotechnology Governance," 59 ALA. L.
REV. 1323, 1374–5 (2008); Institute of Medicine, supra note 5, at 35.

post-commercialization data gathering and reporting requirements would vary according to the assigned level of concern. If a nanomaterial manufacturer believed a certain nanomaterial was misclassified, the manufacturer could apply for reclassification. This classification proposal would establish an information-forcing mechanism that would provide industry the incentive to develop greater data concerning nanotechnology risks but avoid command-and-control dictates that prescribe exactly how to act or impose unduly burdensome requirements on low-risk activities. Such a system could also provide substantial flexibility to adapt governance to new understandings of risk as greater information develops.

Regulatory agencies should also consider incentives they can provide to industry to promote data gathering and reporting. One option would be to create a fast-track review of applications where the applicant develops and submits data beyond what is required under pertinent statutes or where the applicant commits to post-commercialization data gathering and reporting that is not required. Industry would thus be able to get new technology products to market more rapidly, agencies could conduct the same level of review to achieve adequate protection and more data on the emerging technology would be developed. Great Britain, for example, has instituted a voluntary reporting system with some of these characteristics for those involved in developing newly engineered nanotechnology materials.

3.3.2 Filling Regulatory Gaps

Statutes and regulations, almost by definition, are designed to handle regulatory concerns existing at the time of promulgation. It is not surprising that emerging technologies often exacerbate regulatory gaps or introduce new concerns that create new regulatory voids under existing statutes.

Biotechnology, for example, permits the production of genetically modified plants and animals that could have significant environmental impacts, yet the existing US regulatory system often lacks a role for the EPA to review or oversee such products. The EPA lacks authority to regulate genetically modified fish and most other animals and has no role in the approval or field-testing and widespread planting of genetically modified plants other than those modified to be pest-protected. Perhaps even more troubling, it remains unclear whether any regulatory agency has clear authority over transgenic animals not intended for human food or human biologics production.

The development of nanotechnology has also produced regulatory gaps. The most significant arise from the potential for small volumes or masses of nanotechnology products to pose significant human health or environmental risks. Most health, safety and environmental regulations operate on the basis of volume or mass triggers. Examples include the Toxic Substances Control Act's exemption for chemicals made in quantities of less than 10 000 kilograms and the Resource Conservation and Recovery Act's conditionally exempt small quantity generator status for entities disposing under 100 kilograms of waste per year.[9] The assumption that certain generalized quantities of a hazardous substance are necessary to create health or environmental risks, however, likely does not hold true for nanomaterials. Due to nanoparticles' small size and high surface-area-to-mass, such particles may present unusual toxicity concerns in low quantities.

New technologies – particularly those as revolutionary as biotechnology and nanotechnology – disrupt existing regulatory systems. These disruptions can exacerbate problems with existing systems, but can also provide the social and political pressure necessary to fix such deficiencies. Regulatory agencies must get beyond the hurdles created by scientific uncertainty and bureaucratic and status quo inertia to respond more proactively to these challenges. Closing regulatory gaps expeditiously can provide certainty for industry and comfort for the public.

3.3.3 Industry Stewardship

Developing incentives for industry to act in a socially responsible manner can advance many of the emerging technology governance goals discussed above. Such incentives can include economic, public relations, social values and legal mechanisms.

The largest companies generally have strong incentives to maintain robust public confidence in their technological field across the board. These companies have the largest economic stake in a particular technology industry and will be harmed the most by any perceived adverse event, whether traced to their company or to another in the same industry. Confirming this, leadership at some of the nation's largest firms engaged in nanotechnology research and development have been relatively sympathetic to some nanotechnology regulation, due in part to a desire to ensure that consumer reaction to nanotechnology does not follow the same path as consumer concerns about agricultural biotechnology.

---

[9] 40 C.F.R. § 723.50; 40 C.F.R. § 261.5.

Larger technology companies thus have incentives to develop – in concert with government, scientists and public interest groups – guidelines for best management practices concerning emerging technologies. The larger companies also have incentives to exert industry pressure on smaller startups to comply with these best management practices. As noted, any concern or accident related to a particular technology could ripple throughout the entire industry.

In one substantial confirmatory example, DuPont teamed with Environmental Defense to develop a risk management framework for nanomaterials.[10] This framework was designed to "establish a process for ensuring the responsible development of nanoscale materials, which can then be widely used by companies and other organizations." There have also been non-regulatory efforts to create and implement nanotechnology best practices at the international level, such as the joint effort of a number of private and government-funded groups to develop a "Code of Conduct for Responsible Nanotechnology."[11]

That being said, the Environmental Defense-DuPont framework has been criticized by some environmental and labor groups as an attempt by "industry and its allies" to usurp government oversight and public participation in the regulatory process.[12] Also of concern is the potential for larger companies to promote overly expensive or complex management practices in an effort to create barriers to entry for smaller competitors. On the other hand, smaller companies sometimes may be willing to take greater risks than are generally socially acceptable.

There is no question that private, industry-produced best management practices must be appropriately scrutinized, but such efforts should also be encouraged for their potential to produce efficient and rapid governance. Once best management practices are developed, all companies will be heavily incentivized to implement them due to industry peer pressure, public perception and the threat of tort liability if established practices are not observed and an adverse event occurs.

In addition to the public relations and fast-track opportunities identified above, the emerging technology industry can be encouraged to

---

[10]   Nano Risk Framework: A Partnership of Environmental Defense Fund and DuPont (2007), available at http://nanoriskframework.com/ (accessed July 8, 2013).

[11]   Royal Society of the United Kingdom et al., THE RESPONSIBLE NANO CODE (2008), available at http://www.nanotechia.org/activities/responsible-nano-code (accessed July 8, 2013).

[12]   J. Clarence Davies, EPA AND NANOTECHNOLOGY: OVERSIGHT FOR THE 21ST CENTURY 36 (2007).

engage in activities beyond those legally mandated through other incentives, such as potential penalty avoidance. A firm that agrees to conduct regular auditing and self-reporting of regulatory agency-determined practices, for example, could be exempted from certain fines for minor violations that are not intentional or the result of gross negligence. Thus, firms can be incentivized to act beyond what is legally required to address unregulated matters, adopt preventive measures and help regulatory agencies gather greater data.

Under this model, regulatory rules are not intended to set the ideal standard for behavior, but serve as a mandatory back-stop that applies only if firms do not achieve greater protection with alternative arrangements. In this manner, greater protection than mandated can be accomplished at a lower cost to both taxpayers and industry by offering industry the flexibility to figure out how to achieve more efficient protection and by highlighting the importance of public confidence in emerging technology development.

A broad system of industry stewardship could also have substantial long-run returns. Such a system could foster more of an industry ethic of responsibility and a goal of teamwork among the government, industry and consumer organizations. In turn, this teamwork could build commitment among various stakeholders to the governance structure and to cooperation itself, instead of each entity constantly challenging the program and each other over every perceived deficiency.

3.3.4 Agency Expertise and Coordination

Emerging technologies often exacerbate enduring problems with regulatory agency staffing, funding, lack of scientific expertise and coordination. For example, the EPA, FDA, US Department of Agriculture (USDA), and Occupational Safety and Health Administration have each been identified as understaffed, underfunded and lacking properly trained personnel to handle pertinent emerging technologies.

The problems of agency inexperience will be great for many emerging technologies due to the technological complexity and issues arising at the leading edge of scientific knowledge, but it may be particularly severe for nanotechnology and synthetic biology. Both synthetic biology and nanotechnology represent strikingly interdisciplinary fields. In nanotechnology, depending on the particular technology or product, advanced understanding in materials science, chemistry, physics and biology all may be required to analyze risks. Synthetic biology operates at the intersection of various biology, chemistry and engineering disciplines. There are few scientists with sufficient training in the multiple areas for either field, let alone those who work for government agencies. This problem is exacerbated by the disparity between the remuneration such scientists can receive in private technology sectors versus their opportunities at government agencies. Even where the government attracts appropriately educated and trained scientists, once such scientists step out of private or university research, it can become harder to stay at the forefront of their technological field.

Emerging technologies also often raise particular challenges for inter-agency coordination. For example, the regulation of genetically modified plants and animals in the United States implicates as many as a dozen different statutes and six different federal agencies and services. The multiplicity of statutes and agencies has created confusion among regulated industry and the public, reduced clarity regarding scientific standards and requirements, and retarded the efficiency of biotechnology development and regulation. There have even been instances of inconsistent regulations among the FDA, EPA and USDA. Regulatory agency coordination for nanotechnology has been identified as a critical need as well.

Regulatory coordination and consistency for emerging technologies is important on a number of fronts. First, coordination can offer significant cost savings. In a system where agencies are understaffed and underfunded, coordination allows a pooling of personnel, data and other resources, rather than wasteful duplication. Second, because scientific uncertainty rates as one of the most significant problems facing emerging technology regulation, coordinating research concerning human health and environmental risks can stretch scarce agency research resources. Third, a coordinated approach to regulation and requirements can provide efficiency benefits for both government and industry. It is true that agencies operate with different cultures and organizations, but figuring out how to better integrate efforts on emerging technology governance can yield benefits across the board. New governance systems for emerging technologies should include a focus on promoting both intra- and inter-agency coordination.

3.3.5 Governance Adaptability

Emerging technologies develop rapidly. It is often impossible to predict what products and risks will need to be governed even a short time into the future. For example, the United States' Coordinated Framework for Regulation of Biotechnology was adopted in 1986, yet did not cover the regulation of transgenic pest-protected plants. This occurred despite the fact that such products began field-testing only one year later and are now one of the dominant biotechnology products. It is necessary that any emerging technology governance system be flexible enough to adapt, as best as possible, to technological change and advances in scientific understanding of the technology.

Problematically, most regulatory requirements become stringently fixed once put into place and resist attempts at evolution, even in the face of strong evidence that there are significant problems with the existing structure or standards. For example, despite problems that had been identified for many years, the USDA only relatively recently proposed changes to their program governing the import, movement and environmental release of genetically modified organisms. These changes represented the first comprehensive review and revision of the regulations since they were first promulgated in 1987, at the time of the first field trials of genetically modified plants. As noted above, such plants now represent dominant sources of soybeans, cotton and corn in the United States. One purpose of the proposed changes was to bring the USDA regulations in line with the Plant Protection Act, a statute enacted about a decade earlier.

One method for achieving adaptability and flexibility is for emerging technology governance to include mechanisms that allow for incremental changes in governance as needs arise. Such an approach simultaneously provides flexibility in governance and limits the likelihood of quickly upsetting settled expectations for industry. Emerging technology governance should be an iterative process at early stages of technological development and commercialization. A particular system of governance should be developed, followed by data gathering and evaluation, followed by modifications to the system as warranted, in a continuing cycle until scientific understanding and regulation has matured.

Though there are institutional and legal hurdles to establishing such an iterative process,[13] some of these may be overcome by building options into regulatory rules. In addition, instituting a somewhat standardized process for modification allows such change to become part of the expected governance system and should allow future change to occur faster and at lower cost. As more scientific evidence becomes available, this will allow the system to adapt more rapidly.

Governmental agencies should also work with firms to permit flexibility in how regulatory requirements are satisfied to the extent practicable

---

[13]   The Administrative Procedure Act, for example, mandates that final rules cannot be revised once promulgated, except by new rulemaking. 5 U.S.C. §§ 551–9.

while still protecting human health and the environment. Flexibility will allow industry to experiment with economic or technical feasibility and various control approaches, while still ensuring adequate protection. Such experimentation may also help develop additional information on technology risks and the relative advantages of various governance approaches.

3.3.6   Stakeholder Involvement

Critical to these proposals for emerging technology governance is widespread and diverse stakeholder involvement. This involvement should include regular communication with and workshops among a variety of stakeholders, including regulatory agencies, industry representatives, research scientists, environmental organizations, public interest groups, academics and others. Broad stakeholder outreach and dialogue can bring credibility, new ideas, current information, continual feedback and public trust to a governance system.

The communication should include information on the known and unknown risks and benefits of an emerging technology (provided in a form accessible to a broad cross-section of the population), disclosure of new scientific information concerning the technology as it arises, and means for and encouragement of further public involvement. Such communication is particularly important at the early stages of a technology's development because of the public's limited knowledge and awareness of the technology. A well-informed public, in turn, can allow consumers to "vote with their dollars" to try to influence industry decisions. Of course, there must also be a high level of transparency in regulatory decision-making and activity.

The communication efforts should include specialized outreach to smaller technology companies. Current health and environmental regulatory programs generally evolved around existing, mature industries at times when there were relatively fewer and larger companies. Larger companies are generally more aware of and able to respond to regulatory requirements. On the other hand, startup companies and small firms, including many that are not familiar or sophisticated with respect to existing health and environmental regulations, will need to be made aware of regulatory requirements. In some cases, such companies may need assistance with responding to and complying with regulations. Training and technical assistance on compliance for startups and small companies could be provided.

As noted, public trust in an emerging technology and its governance is critical to the success of the technology. The failure to provide for adequate stakeholder involvement and public communication, in particular, has been identified by some as one reason for some of the public backlash against biotechnology. The potential for a public reaction against emerging technologies is elevated by the complex science involved, the high level of uncertainty concerning risk and the potential for interest group polarization. Perhaps recognizing this concern, there are growing efforts to more proactively incorporate discussion of ethical, legal and social implications of synthetic biology into its research and development in the United States and internationally.

Communication will not resolve all concerns or potential for conflict, but can go a long way toward establishing broad public trust. Implementing these informational measures in concert with other revisions recommended above can produce a framework for emerging technology governance that could simultaneously better protect against health and environmental risk, develop greater information, permit rapid technological advancements and maintain public confidence in the governance system.

3.4 CONCLUSION

The new governance proposals outlined here, and their potential for success, are undoubtedly optimistic. Emerging technology governance presents many complex social, political, cultural and technological issues, some of which have been only limitedly discussed. Implementing these recommendations would be neither simple nor frictionless. These proposals are meant, however, to provide a road map in the form of a best-case argument for what could be achieved.

Studies of public perception of various emerging technologies, including those discussed here, have found that individual perceptions of the technologies tend to polarize along traditional cultural and social lines. The groups that polarize are not the same for each technology, but each technology faces significant risk of producing concerning cultural schisms. This polarization implicates not only the perceived scientific benefits and risks of the technology in question, but also perceptions of the potential economic and social effects of a given technology.

This public perception highlights that the development of any emerging technology and the system for governing it are inevitably and dynamically intertwined. Each will continually affect the other. The new governance proposals developed here aim to institute a collaborative, transparent and adaptable system at an early stage of technological development to ameliorate the potential for social division over a technology and to set the tone for a long-term, collaborative governance model.

Given the uncertainty surrounding an emerging technology's development and risks, there will be inherent limitations concerning how specific a framework can be developed at early stages. The proposal outlined above is not intended to be exhaustive by any means, but to develop a core structure that will provide greater certainty for the public and industry and allow details to be developed as knowledge evolves.

Though the early stages of a technology's development can be a challenging time to develop a governance framework, as there are still many unknowns, it can also be an opportune moment to take advantage of the flexibility of a new approach. At the early stages, fewer interests have vested around particular governance regimes, there are not as significant sunk costs to overcome, and industry and the public are less wed to a status quo. The early stages of an emerging technology's development present a unique opportunity to shape its future. But it is an opportunity that does not remain open forever. Interests, investment and opinion can quickly vest around regulatory and governance expectations. It is important to put an appropriate governance system in place early in a technology's developmental stages and before the commitment to the status quo becomes too great.

For the first time in history, there is the opportunity for governance systems to develop simultaneously with technologies, permitting proactive rather than reactive management structures. The opportunity to reap the potentially spectacular health, environmental and economic benefits of emerging technologies are great, but these opportunities will be severely hampered if the technologies are not managed properly. The opportunities will be hampered because society will face inefficient costs and delays in technological development and unnecessary technological risks. In addition, inadequate regulation can lead to distrust of the governance system and to high-profile problems, each of which can result in public backlash against the technology. The emerging technology governance system proposed here offers a framework to navigate these hazards in an effort to promote responsible and valuable technology development.

4. An integrated framework for governing emerging technologies such as nanotechnology and synthetic biology

LeRoy Paddock and Molly Masterton

4.1 INTRODUCTION

The continued growth of new technologies such as nanotechnology and synthetic biology poses an array of environmental and public health benefits and risks. The complexity of the technologies and the speed with which they are entering the marketplace require a much more sophisticated form of governance than has been commonly employed to deal with environmental and health risks in the past. This new system should include what some authors have called an "experimentalist" form of regulation integrated with a wide variety of other tools including information disclosure, public engagement, the continuing possibility of tort liability, and a range of self-governance approaches. This chapter builds on earlier recommendations for an integrated governance system for nanotechnologies and extends this analysis to synthetic biology. In particular, this chapter focuses on the need for a more adaptive system of governance that allows government stakeholders to learn as experimentation proceeds and rapidly adjust risk management approaches based on what has been learned, coupled with other governance approaches such as information disclosure, codes of conduct, liability, and public dialogue.

4.2 INDUSTRY GROWTH UPDATES

4.2.1 Nanotechnology

Nanotechnology has become one of the defining technologies of the twenty-first century. As of February 2013, the Project on Emerging

Technologies indicated that the numbers had grown significantly, with 1316 nano-related products, produced by 587 companies in 30 countries.[1] These technologies can be found in products as different as sunscreen and the fuselage of the new Boeing 787. The substances most commonly used in the industry continue to be metals, carbon fibers, and silicon/ silica. According to the Project on Emerging Technologies, "[t]he most common material mentioned in … product descriptions is now silver (313 products). Carbon, which includes fullerenes, is the second most referenced (91), followed by titanium (including titanium dioxide) (59), silica (43), zinc (including zinc oxide) (31) and gold (28)."[2]

Through its applications in the fields of medicine, energy, and environmental improvement nanotechnology may provide great benefits to society. The nanoscale is the level at which many fundamental life processes occur, and as such, nanotechnology may prove to be an unprecedented resource for fighting diseases.[3] In the energy industry nanotechnology can be used to more efficiently capture solar energy.[4] The use of nanosilvers in pesticides has become particularly common because it allows lower-quantity use, timed release of applications, and may better target the characteristics of certain pests.[5,6] The United States produces the most products followed by Europe and East Asia. The

---

[1]   The Project on Emerging Technologies, Nanotechnology Consumer Products Inventory, available at http://www.nanotechproject.org/inventories/consumer/ (accessed July 12, 2013). The Project on Emerging Technologies is a partnership between the Woodrow Wilson International Center for Scholars and the Pew Charitable Trusts.

[2]   The Project on Emerging Nanotechnologies, Analysis/Consumer Products, available at http://www.nanotechproject.org/inventories/consumer/analysis_ draft/ (accessed July 12, 2013).

[3]   Lindsay V. Dennis, Nanotechnology: unique science requires unique solutions, 25 Temp. J. Sci. Tech. & Envtl. L. 87, 94 (2006).

[4]   US Environmental Protection Agency, Office of the Science Advisor, Nanotechnology White Paper 27 (2007), available at http://www.epa.gov/OSA/ pdfs/nanotech/epa-nanotechnology-whitepaper-0207.pdf   (accessed   July   12, 2013).

[5]   US Environmental Protection Agency, Regulating Pesticides that Use Nanotechnology,   available   at   http://www.epa.gov/pesticides/regulating/nano technology.html (accessed July 12, 2013).

[6]   Nanosilvers used in pesticides are currently subject to EPA regulations under FIFRA, see infra notes 104–5 and accompanying text.

market for scientific production in nanotechnologies continues to advance at a rate of 10 percent a year.[7]

The unique properties that enable nanotechnology to be used in many different products are also the source of significant risks, particularly in the environmental and health contexts. Studies during the initial wave of products entering the market have focused on the small size of nanomaterials, which may give them the ability to enter human cells and alter biological processes at the cellular level.[8] Studies on nanosilvers have indicated that they may have negative antimicrobial effects on ecosystems. Nanoparticles such as titanium dioxide and silver may be ecotoxic, raising the concern that "[t]he antifungal and antibacterial activity of manufactured nanoparticles may seriously threaten free-living nitrogen-fixing bacteria and symbiotic relationships involving fungi, bacteria, and plants."[9] Another more notorious risk was introduced when scientists found that carbon nanotubes may cause lung damage similar to that caused by asbestos.[10] Further, a recent study showing that nanomaterials can biomagnify in a simple food chain has raised concerns about the use of nanomaterials for food packaging as well as more general ecological impacts.[11] The real-life impacts of many nanomaterials on human bodies remain uncertain, however, making traditional governance and risk assessment tools less useful.[12] As

---

[7] Vincent Mangematin and Steve Walsh, THE FUTURE OF NANOTECH-NOLOGIES 4 (Working Paper No. 1, 2012), available at http://hal.grenoble-em.com/docs/00/65/80/34/PDF/future_of_Nanotechnologies.pdf (accessed July 12, 2013).

[8] National Research Council, A MATTER OF SIZE: TRIENNIAL REVIEW OF THE NATIONAL NANOTECHNOLOGY INITIATIVE 78–79 (2006), available at http://books.nap.edu/openbook.php?record_id=11752 and page=R1 (accessed July 12, 2013).

[9] International Risk Governance Council, NANOTECHNOLOGY RISK GOVERNANCE: RECOMMENDATIONS FOR A GLOBAL, COORDINATED APPROACH TO THE GOVERNANCE OF POTENTIAL RISKS 8 (2007), available at http://www.irgc.org/IMG/pdf/PB_nanoFINAL2_2_.pdf (accessed July 12, 2013).

[10] Maria Lee, "Risk and Beyond: EU Regulation of Nanotechnology," 35 EUR. L. REV. 799, 801 (2010).

[11] See Werlin et al., "Biomagnification of Cadmium Selenide Quantum Dots in a Simple Experimental Microbial Food Chain," 6 NATURE NANO-TECHNOLOGY 65–71 (2011).

[12] For more information on recent environmental, health and safety-related studies on nanotechnology, see Research Inventory, THE PROJECT ON EMERGING NANOTECHNOLOGY, available at http://www.nanotechproject.org/inventories/ehs/ (accessed July 12, 2013).

described by Professor Maria Lee of the University College in London, "normative debate continues on around the acceptability of any risk as compared with benefits, together with the level of certainty required to take action."[13] The prevalence of risks makes apparent the need for additional research, but also presents great challenges for the current governance of the development and distribution of nanomaterials. We advocate an approach that combines traditional regulation with various governance tools in order to most effectively retain the societal benefits of nanotechnologies while managing their possible risks.

4.2.2 The Emergence of Synthetic Biology

Synthetic biology is another emerging technology that has grown at a rapid pace with little regulation during its developmental phase. It gained recognition as a field no earlier than 2004 when the first international synthetic biology conference, entitled "SB 1.0," was held.[14] Synthetic biology can be most accurately described as sitting at the cross-section of engineering and biology, where it seeks to build biological systems "from scratch" by allowing scientists to create systems that are more efficient and targeted for certain levels and timing of production.[15] In other words, it allows scientists to more easily engineer biology.[16] One example of this is found in synthetic genomics and DNA-based device construction, where rather than trading genes between species as in traditional genetic engineering, synthetic biology allows scientists to write an entirely new genetic code on a computer and makes it available to be downloaded or "printed" as needed to manufacture projects.[17]

This new idea of building living structures from the ground up has opened opportunities both to advance basic knowledge of life forms and to create new and useful products. By applying engineering principles to living systems, synthetic biology is a sort of hybrid of many different disciplines, including engineering, biology, chemistry, computer science,

---

[13]   Lee, supra note 10, at 801.

[14]   Joy Y. Zhang, Claire Marris and Nikolas Rose, "The Transnational Governance of Synthetic Biology: Scientific Uncertainty, Cross-Borderness and the 'Art' of Governance," 5 (BIOS, The London School of Economics and Political Science, Working Paper No. 4, 2011), available at http://royalsociety.org/uploadedFiles/Royal_Society_Content/policy/publications/2011/4294977685.pdf (accessed July 12, 2013).

[15]   Drew Endy, "Foundations for Engineering Biology," 438 NATURE 449, 453 (2005).

[16]   Zhang et al., supra note 14, at 3.

[17]   Id.

information technology, and even nanotechnology.[18] Research on proto-cells and the creation of new bacteria, in concert with synthetic genome projects, presents the potential for a "mini-factory producing various substances, from treatments for diseases to weapons of terror."[19] The United States government is at the head of industrial research in synthetic biology. In 2010 there were approximately 184 US institutions engaging in synthetic biology-related work.[20] Between 2005 and 2010 the government invested an estimated $430 million in synthetic-biology research, with the Department of Energy funding the majority of the research.[21] The UK, by comparison, invested between $30 and $53 million in synthetic biology-related research during this timeframe.[22]

Although estimations of current market value vary greatly, some have forecasted a market value of over $3.5 billion within the next decade.[23] An initial inventory by the Woodrow Wilson Center identified over 60 already-developed products in the US that will be entering the market in either the short- or long-term.[24] The products within sight include biofuels and other energy-related materials, drugs, vaccines, chemicals, cosmetics, foods, and explosives.[25] Many of the potential uses for this technology would be helpful to the environment, including innovative methods for cleaning up oil spills and other environmental disasters, carbon sequestration, creation of new energy sources and more environmentally-sound biodegradable products. One of the most-discussed energy applications of synthetic biology in the environmental

---

[18]   Id. at 9.

[19]   Synthetic Biology Project, ETHICAL ISSUES IN SYNTHETIC BIOLOGY: AN OVERVIEW OF THE DEBATES, 7 (2009), available at http://www.synbioproject.org/library/publications/archive/synbio3/ (accessed July 12, 2013).

[20]   Synthetic Biology Project, TRENDS IN SYNTHETIC BIOLOGY RESEARCH FUNDING IN THE UNITED STATES AND EUROPE (2010), available at http://www.synbioproject.org/library/publications/archive/research funding/ (accessed July 12, 2013).

[21]   Id.

[22]   Id.

[23]   Zhang et al., supra note 14, at 6.

[24]   See Synthetic Biology Project, INVENTORY OF SYNTHETIC BIOLOGY PRODUCTS – EXISTING AND POSSIBLE (July 27, 2012), avail-able at http://www.synbioproject.org/process/assets/files/6631/_draft/synbio_applications_wwics.pdf (accessed July 12, 2013).

[25]   Id.

field is the creation of biofuel-producing algae[26] and consolidated bio-processing of cellulose to create ethanol (a process known in the industry as "CBP").[27] The technology also holds many promises for advances in medicine. Engineers at the Synthetic Biology Engineering Research Center (SynBERC), for example, have worked to engineer a bacterium with the potential to destroy tumors.[28] Biosensors, which use engineered bacteria to detect water and soil contaminants such as arsenic, present another application on the horizon with broad societal benefit.[29] The growth of the synthetic biology industry is not limited to industry and academic research-based laboratories. As noted by experts at the J. Craig Venter Institute:

> the ability to carry out DNA synthesis is no longer confined to an elite group of scientists as was the case for the first several decades of research using recombinant DNA. Now, anyone with a laptop computer can access public DNA sequence databases via the Internet, access free DNA design software, and place an order for synthesized DNA for delivery.[30]

The phenomenon of do-it-yourself synthetic biology, which some have termed "garage" synthetic biology,[31] is one of several unique character-istics that place the field of synthetic biology in the open-source development arena.

The same technologies that may present such great advantages, how-ever, are also capable of creating human and environmental hazards. Two main concerns with synthetic biology are biosafety and biosecurity. There

---

[26]   See generally Michael S. Ferry, "Synthetic Biology Approaches to Biofuel Production," 3 BIOFUELS 9–12 (2012), available at http://biodynamics.ucsd.edu/pubs/articles/Ferry12.pdf (accessed July 12, 2013).

[27]   Michele S. Garfinkel, Drew Endy, Gerald L. Epstein and Robert M. Friedman, J. Craig Venter Institute, SYNTHETIC GENOMICS: OPTIONS FOR GOVERNANCE 12 (2007), available at http://www.jcvi.org/cms/research/projects/syngen-options/ (accessed July 12, 2013).

[28]   SynBERC, SYNTHETIC BIOLOGY ENGINEERING RESEARCH CENTER, available at http://www.synberc.org/ (accessed July 12, 2013).

[29]   See Royal Academy of Engineering, SYNTHETIC BIOLOGY: SCOPE, APPLICATIONS AND IMPLICATIONS 8, 38 (2009), available at http://www.raeng.org.uk/news/publications/list/reports/Synthetic_biology.pdf (accessed July 12, 2013).

[30]   Garfinkel et al., supra note 27, at 6.

[31]   See generally Heidi Ledford, "Garage Biotech: Life Hackers," 467 NATURE 650 (2010).

is a potential for unintended release of synthetic organisms to accidentally cause epidemics, as well as a risk that bioterrorists could intentionally use this new technology to adjust existing viruses or create new and dangerous viruses.[32] The growing prevalence of DIY synthetic biology, of course, adds to the concern. Further, some have speculated that new organisms may be able to self-replicate beyond our control.[33] This concept has led to a central environmental concern, which is that self-propagating organisms could disrupt natural environments through competition or by spreading genetic material across species boundaries.[34]

There is good reason for treating synthetic biology together with nanotechnology in the context of designing a policy framework. Foremost, as the Woodrow Wilson Center's Synthetic Biology Project has noted, "insofar as synthetic biology occurs at the nanoscale, it would not be unreasonable to consider it a form of nanotechnology."[35] There are also many interactions between the two developing fields. Some applications of synthetic biology may provide the "production facilities" for nanoscale fabrication.[36] Due to the interconnectedness of the fields, they have often been included in the larger, converging group known as "emerging technologies."[37] One commonality of almost all emerging technologies through history, including, for example, genetic engineering and robotics, is the fundamental "newness" and uncertainty of the risks.[38] Nanotechnology and synthetic biology also share commonalities in the realms of environmental and health hazards. Further, the diversity of the

---

[32]   See The European Commission, "Synthetic Biology: Key Policy Issues and Options," NEWS ALERT ISSUE 216 (2010), available at http://ec.europa.eu/environment/integration/research/newsalert/pdf/216na1.pdf (accessed July 12, 2013).

[33]   See International Risk Governance Council, RISK GOVERNANCE OF SYNTHETIC BIOLOGY 5 (2009), available at http://www.irgc.org/IMG/pdf/IRGC_Concept_Note_Synthetic_Biology_191009_FINAL.pdf (accessed July 12, 2013).

[34]   See id. at 12.

[35]   Synthetic Biology Project, supra note 19, at 10.

[36]   Philip Ball, "Synthetic Biology for Nanotechnology," 2005 NANO-TECHNOLOGY 16 (2005).

[37]   See generally CONVERGING TECHNOLOGIES FOR IMPROVING HUMAN PERFORMANCE: NANOTECHNOLOGY, BIOTECHNOLOGY, INFORMATION TECHNOLOGY AND COGNITIVE SCIENCE (Mihail C. Roco and William Sims Bainbridge, eds., 2004), available at http://www.wtec.org/ConvergingTechnologies/Report/NBIC_report.pdf, accessed July 12, 2013.

[38]   See Zhang et al., supra note 14, at 3 (noting that in the case of emerging technologies such as synthetic biology, many of the future implications "are not only difficult to predict but are also fundamentally unknowable").

technologies and risks involved make governance difficult for both; as the International Risk Governance Council (IRGC) points out, "in no country is there a single regulatory structure that covers food, chemicals, personal care products, medical devices, water quality, and so on."[39] Although many social discussions of risks and benefits have separated nanotechnology and synthetic biology, it may be wise to develop a general governance approach that meets the common concerns of the two industries.

4.3   INTEGRATED APPROACH TO GOVERNING RAPIDLY ADVANCING TECHNOLOGY

4.3.1   Overview of Elements and Concepts

The principal challenges presented by the rapid emergence of nanotechnology and, now synthetic biology, include:

1   the speed at which the industry is developing and the expected scale of the industry;
2   the diversity of technologies involved;
3   the low levels of public awareness about nanotechnologies;
4   the race to be the first to the market with new products in a competitive global economy;
5   the potential extraordinary social benefits of some nanotechnologies;
6   the nature of the risks that may be created by some nanotechnologies; and
7   the ethical, legal and social issues associated with some of the technologies.[40]

In the context of the rapid development and deployment of nanotechnologies over the past decade, we have previously recommended an integrated governance approach, anchored in government regulation but also incorporating economic and value-based behavioral drivers. A similar approach is needed for synthetic biology. Such an integrated approach would make adaptive regulations part of a broader system of governance approaches, and would utilize governance tools to encourage both

---

[39]   International Risk Governance Council, supra note 9, at 8.
[40]   LeRoy Paddock, "An Integrated Approach to Nanotechnology Governance," 28 UCLA J. ENVTL. L. & POL'Y 251, 255 (2010).

organizations and individuals to take responsibility in the context of nanotechnology. The approach should include:

- advanced, adaptive regulations that regularly generate new information on how the governance system is functioning and that allows stakeholders to quickly respond the new information;
- new sources of information including data on publically accessible websites, product labels that facilitate consumer decision making, and data disclosures that allow more informed government decision making;
- codes of conduct and industry standards that help anticipate problems and respond more quickly than government has traditionally been able to respond;
- new forms of public dialogue that engage the general public in the governance discussion including discussions of research priorities and risk research;
- a better understanding of the factors that drive businesses to themselves be cautious about deployment of new technologies such as reputation and investor relations, insurance availability, and consumer pressure, among other factors;
- retention of common law liability regimes that can temper hasty decisions to bring potentially risky products to the market and that can spur research organizations and manufactures to protect the safety of their workers;
- voluntary government programs that encourage industry to take a leadership role in risk management beyond the minimums required by law.[41]

Some groups have advocated for a simple strategy based on the precautionary principle to deal with potentially high-risk materials. For example, Friends of the Earth and 57 other organizations have recently commented on how what they call "extreme genetic engineering" should be governed. Rejecting the premise that existing EPA regulations will be sufficient, their "Principles for the Oversight of Synthetic Biology" instead recommend the use of the precautionary principle and the creation of mandatory synthetic biology-specific regulations.[42] They

---

[41]   Id. at 266.
[42]   See Friends of the Earth and The International Center for Technology Assessment, THE PRINCIPLES FOR THE OVERSIGHT OF SYNTHETIC BIOLOGY (2012), available at http://www.wilsoncenter.org/sites/default/files/principles_for_the_oversight_of_synthetic_biology.pdf (accessed July 12, 2013).

called for "a moratorium on the release and commercial use of synthetic organisms" until a thorough study of all the environmental and socio-economic impacts of this emerging technology has taken place.[43] However, such a strategy may be problematic considering how far technology has already come in the analogous field of nanotechnology:

> The globally competitive nature of nanotechnology development suggests the need for a multifaceted governance system, since regulatory activity in one country (such as a ban based on a precautionary approach) may simply push the development of the technology to another country. Rather, governance mechanisms will have to generate new information, engage the public and spur action (both regulatory and self-directed) wherever the technology is being developed and used.[44]

4.3.2   An Adaptive Regulatory System

Regulations for emerging technologies require flexibility; as observed by the Project on Emerging Technologies in its study *Managing the Effects of Nanotechnology*:

> The rapid development of [nanotechnology] also means that government managers always will be operating with outdated information, and data on [nanotechnology] effects will lag behind commercial applications. Priorities for research and for regulation will need to shift constantly. We have moved into a world which is, as David Rajeski states, "dominated by rapid improvements in products, processes, and organizations, all moving at rates that exceed the ability of our traditional governing institutions to adapt or shape outcomes."[45]

An adaptive management approach has been suggested as one method of overcoming the lack of information regarding many technologies without either halting technology development or risking the introduction of technologies that may have adverse environmental or health impacts. In other words, "nanoregulation must be regarded as a dynamic affair which must adapt to the evolution of scientific knowledge and applications and

---

[43]   Id. at 3.
[44]   Paddock, supra note 40, at 261.
[45]   J. Clarence Davies, MANAGING THE EFFECTS OF NANO-TECHNOLOGY 2 (2006), available at http://www.nanotechproject.org/process/assets/files/2708/30_pen2_mngeffects.pdf (accessed July 12, 2013).

public attitude. A continuous updating must be part of the governance of nanotechnology."[46]

The International Risk Governance Council defines risk governance as a discipline that "applies the principles of good governance to the identification, assessment, management and communication of risks. It incorporates such criteria as accountability, participation, and transparency within the procedures and structures by which risk-related decisions are made and implemented."[47] Although the potential benefits of emerging nano and biotechnologies are great and reach across many fields, the potential risks combined with a general lack of information make risk governance a priority. As the IRGC Framing Nano report noted, there is "a remarkable lack of information on some key aspects of concerning the environmental impacts of manufactured nanoparticles, which currently prevents a better understanding and assessment of the toxicity and ecotoxicity of manufactured nanoparticles to the key ecosystem organisms."[48]

Gregory Mandel and other commentators have recommended a new governance model that focuses on the early stages of a technology's development and allows for governance to change with the technology by creating "a more proactive, flexible form of governance; a governance process rather than intractable regulatory rules."[49]

> The combination of common concern about uncertainty and lack of a status quo can create a unique window of opportunity for a broadly-developed and widely-supported new governance model. Such a model, due to its wider base of support, could be instituted far earlier in a technology's development, providing assurance to the public and stability for industry. This window, however, will not remain open indefinitely. As particular regulatory decisions are made or relied upon, and investment is made in particular products and uses, many stakeholders will become less flexible.[50]

---

[46] Elvio Mantovani, Christoph Meili, Andrea Porcari and Markus Widmer, FRAMING NANO, MAPPING STUDY ON REGULATION AND GOVERNANCE OF NANOTECHNOLOGIES 8 (January 2009), available at http://www.framingnano.eu/images/stories/FramingNanoMappingStudyFinal.pdf (accessed July 12, 2013).

[47] International Risk Governance Council, WHAT IS RISK GOVERNANCE?, available at http://www.irgc.org/risk-governance/what-is-risk-governance/ (accessed July 12, 2013).

[48] International Risk Governance Council, supra note 9, at 32.

[49] Gregory N. Mandel, "Regulating Emerging Technologies," 1 LAW, INNOVATION AND TECH. 75, 75 (2009).

[50] Id. at 81.

In light of the complexities, agencies could seek to create a risk classification structure based on currently available information. Such a system, as Mandel suggests, might classify various nanomaterials "as having negligible, low, medium, or high concern" and could incentivize industries to report by "fast-tracking" review of products for those who satisfy or go beyond what is required.[51] While this is only one mechanism for developing risk governance, Mandel argues that any a system for emerging technology governance should take place at the earliest stage possible, "followed by data gathering, followed by result evaluation, followed by modifications to the system as warranted, in a continuing cycle until industry and scientific understanding has matured."[52]

A related theory is the relatively new concept of "experimentalism" in governance.[53] An experimentalist approach, which may begin with the national government funding of state- or industry-based regulatory monitoring activities, is well-suited to the uncertain and constantly shifting realm of emerging technologies. According to Charles Sabel and William Simon, experimentalism works by allowing local units (for example, companies) broad discretion to meet governance goals in whichever manner they see fit but requires them to report results, comment on their successes and failures, and participate in a robust peer review process.[54] The result is a pool of possible regulatory routes that may be compared and utilized or thrown out based on their relative successes. Importantly, the national "framework goals, performance measures, and decision making procedures themselves are periodically revised on the basis of alternatives reported and evaluated in peer reviews, and the cycle repeats."[55] Such an experimentalist method was used in the Food and Safety Modernization Act of 2010, which gave food processors flexibility to meet safety requirements by developing their own Hazard Analysis and Preventative Control Plans and allowed the FDA to adjust its regulatory actions based on the results of each facility's plan.[56]

An experimentalist, risk-based governance structure is consistent with the developing agreement amongst regulatory scholars that emerging technologies will require an "incremental, reflexive, and cooperative"

---

[51]   Id. at 84.
[52]   Id. at 89.
[53]   See generally Charles F. Sabel and William H. Simon, "Minimalism and Experimentalism in the Administrative State," 100 GEO. L.J. 53 (2011).
[54]   See id. at 79.
[55]   Id.
[56]   Id. at 85.

approach.[57] In the UK, commentators as well as regulatory bodies have largely embraced risk-based regulation. Julia Black and Robert Baldwin describe it as ideal for creating a system highly responsive to change in products regulated, scientific understanding of risks, and shifting public perceptions. Their "really responsive" approach requires that regulators pay close attention to five key factors:

1   the behavior, attitudes, and culture of regulatory actors;
2   the institutional setting of the regulatory regime;
3   the different logics of regulatory tools and strategies (and how these interact);
4   the regime's own performance over time; and, finally,
5   changes in each of these elements.[58]

Experimentalist regimes, in particular, may be best suited for the transnational development that is already visible in the fields of nano-technology and synthetic biology; the wider the experimental pool, the greater the chance of finding the method(s) to advance common goals.[59]

An experimentalist approach may not require new legislation. Professor Leon Fuerth argues that what he calls "anticipatory governance" may be based on existing Presidential authority to execute federal law. He notes that this authority provides a basis to "mobilize the full capacities of government, and speed up the process of detecting error and propagating success."[60] Fuerth advocates a system of constant sampling and feedback mechanisms between the White House and the agencies, such that a flexible, "constant reassessment and recalibration of policies" can occur.[61]

---

[57]   Gary E. Marchant, Douglas J. Sylvester and Kenneth W. Abbott, "Risk Management Principles for Nanotechnology," 2 NANOETHICS 43 (2008).

[58]   Julia Black and Robert Baldwin, "Really Responsive Risk-Based Regulation," 32 LAW & POL'Y 181 (2010).

[59]   See Charles F. Sabel and Jonathan Zeitlin, EXPERIMENTALISM IN TRANSNATIONAL GOVERNANCE: EMERGENT PATHWAYS AND DIFFUSION MECHANISMS 1–2 (2011) (unpublished paper presented at the panel of Global Governance in Transition, Montreal, March 2011).

[60]   Leon Fuerth, "Operationalizing Anticipatory Governance," 2 PRISM 31 (2011), available at http://www.ndu.edu/press/lib/images/prism2-4/Prism_31-46_Fuerth.pdf (accessed July 12, 2013).

[61]   Id. at 138.

4.4 APPLICATION OF GOVERNANCE MODELS TO SYNTHETIC BIOLOGY

As a rapidly emerging technology that sits at the intersection of engineering and biology and presents a variety of possible risks to public health and the environment, synthetic biology has been identified by the IRGC as an area where "there may be significant deficits in risk governance structures and processes."[62] In the United States and elsewhere, however, discussions on governance options appear to still be in their initial phases. A landmark in synthetic biology governance occurred in the US in May of 2010 when President Obama asked the Commission for the Study of Bioethical Issues to undertake "a study of the implications of this scientific milestone ... [and] consider the potential medical, environmental, security, and other benefits of this field of research, as well as any potential health, security, or other risks."[63] This presidential request occurred immediately after the introduction by the J. Craig Venter Institute of the first bacterial cell controlled entirely by a synthetic genome.[64] In its study entitled "New Directions: The Ethics of Synthetic Biology and Emerging Technologies," the commission analyzed governance options. It found that rather than the extreme options of issuing a moratorium or letting science develop unhampered, an ideal "coordinated approach" could be utilized by the Executive Office and relevant agencies to:

1   leverage existing resources by providing ongoing and coordinated review of developments in synthetic biology;
2   ensure that regulatory requirements are consistent and non-contradictory; and
3   periodically and on a timely basis inform the public of its findings.[65]

---

[62]   International Risk Governance Council, Risk Governance of Synthetic Biology, supra note 33, at 3.
[63]   Synthetic Biology Project, PROJECT NEWS: THE GOVERNANCE OF SYNTHETIC BIOLOGY, available at http://www.synbioproject.org/news/project/the_governance_synthetic_biology (accessed July 12, 2013).
[64]   Id.
[65]   See Presidential Commission for the Study of Bioethical Issues, NEW DIRECTIONS: THE ETHICS OF SYNTHETIC BIOLOGY AND EMERGING TECHNOLOGIES 127 (2010), available at http://bioethics.gov/cms/synthetic-biology-report (accessed July 12, 2013).

The commission generally concluded that no major change to the statutory, regulatory, and voluntary governance frameworks would be necessary for review and monitoring of synthetic biology as it develops.[66] Nor did it see a need to create new agencies or oversight bodies to focus specifically on synthetic biology, but it noted that regulatory bodies such as the Office of Science and Technology Policy or the Emerging Technologies Interagency Policy Coordination Committee should work closely with scientists to monitor the field and foster a dialogue with the public. It did not look at synthetic biology as a radically new field, finding instead that "[i]n many ways, synthetic biology is an extension of genetic engineering and part of an increasingly interconnected network of scientific disciplines including, among others, nanotechnology and information technology."[67] This approach appears consistent with both the "experimentalism" approach and Fuerth's belief that the executive agencies do have sufficient authority to adopt an anticipatory approach to synthetic biology governance. It is also encouraging that the need to keep the public informed is recognized by the commission.

The commission did emphasize the "low-probability, possible high-impact nature" of emerging technology risks requires an analysis of possible gaps in current risk assessment practices pertaining to synthetic biology so that an effective process can be established for assessing products prior to field release.[68] Rather than issuing regulations in this early phase, the commission recommended the creation of voluntary guidelines for manufacturers and stressed the importance of maintaining a dialogue with all stakeholders in the development process. This idea links nicely to the idea of integrated governance that we have advocated. The commission's five guiding principles were:

(a) public benefit;
(b) responsible stewardship;
(c) intellectual freedom and responsibility;
(d) democratic deliberation; and
(e) justice and fairness.[69]

---

[66] Id. at 124.
[67] Id.
[68] Id. at 128, 131.
[69] Id. at 4.

Still, according to the Woodrow Wilson Center's Synthetic Biology Project, this framework has not been backed up by mechanisms needed to track progress of the recommendations.[70]

Jennifer Kuzma and Todd Tanji have presented synthetic biology governance options in the form of four possible policy stances: (1) promotional, or encouraging development; (2) permissive, or neutral toward development; (3) precautionary, or slowing down development; and (4) preventative, or blocking development.[71] They stress that no single one of these approaches should be uniformly applied to the synthetic biology industry but rather they should be used as appropriate in different areas of the field depending on the associated risk.[72] The Commission for the Study of Bioethical Issues suggested that risk assessment was needed, focused on voluntary initiatives and did not indicate a desire to slow development, thus making their approach permissive or even promotional. The approach in Europe at the federal level has also ranged from permissive to promotional, in that there is a general consensus that the current regulatory framework for genetically modified products and environmental releases will be sufficient to cover synthetic biology.[73] In order to determine the level of caution that should be taken in our integrated approach, however, it is important to consider the unique governance challenges that will be posed by synthetic biology.

Several key governance challenges have been identified by the BIOS program at the London School of Economics and Political Science. First is the inherent uncertainty of the risk; the authors note that the implications of synthetic biology are "not only difficult to predict but are fundamentally unknowable."[74] BIOS also stresses the "cross-borderness" of synthetic biology.[75] This refers to its span across different disciplines that are usually not regulated together, and it will be a key challenge for governance. Specifically, communicating risks across these boundaries will be a challenge. In the context of biosafety, leading synthetic biologist Drew Endy notes:

---

[70]   See Synthetic Biology Project, SYNTHETIC BIOLOGY SCORECARD, available at http://www.synbioproject.org/scorecard/ (accessed July 12, 2013).

[71]   See Jennifer Kuzma and Todd Tanji, "Unpackaging Synthetic Biology: Identification of Oversight Policy Problems and Options," 4 REG. & GOVERNANCE 92, 92 (2010).

[72]   See id.

[73]   See International Risk Governance Council, supra note 33, at 10.

[74]   Zhang, et al., supra note 14, at 3.

[75]   Id. at 9.

The majority of people coming into synthetic biology aren't biologists. They're physicists or computer scientists or electrical engineers and so they're just of a different culture. They don't have a lot of experience with microbiological safety. So you need to gain access or transmit knowledge across not just a generational gap, but across cultural divides.[76]

Across risk factors including health and environmental safety, even experienced and well-intentioned scientists may fail to recognize risks that lie outside their area of expertise.

BIOS also points to the "transnational" nature of the current synthetic biology industry as a key challenge. In many ways the synthetic biology industry right now mirrors an "open innovation" process. International financial backing and resource exchange are common, one example being technology sharing between young scientists from around the world in the "international genetically engineered machine" (iGEM) competition.[77] The global nature of synthetic biology research calls for transnational cooperation, and as such BIOS stresses the need for strong inter-relations between geopolitical regions, academics disciplines and industrial sectors. This suggests that while national laws may be helpful in managing some of the risks associated with synthetic biology, other tools such as codes of conduct, liability, the possibility of reputational harm, public engagement, and information disclosure may be needed to deal with risks that cross borders in international trade.

Pointing to the unique combination of challenges presented by synthetic biology, including its unknown risks and its span across several disjointed disciplines, BIOS discusses the necessity of what they describe as the "art of governance":[78]

In the light of these three challenges, we argue that scientifically informed, evidence-based approaches to policy-making, while essential, are insufficient. It is time to bring back a sense of the 'art' to the governance of biotechnology: an approach which employs proactive, open-ended regulatory styles able to work with uncertainty and change, to make links across borders, and to adapt to evolving relations among changing stakeholders, including researchers, research funders, industry, and multiple publics.[79]

---

[76]   Filippa Lentzos, "Synthetic Biology in the Social Context: The U.K. Debate to Date," 4 BIOSOCIETIES 303, 319 (2009).

[77]   See generally IGEM, available at http://igem.org/About (accessed July 12, 2013).

[78]   Zhang, et al., supra note 14, at 23.

[79]   Id. at 4.

According to the BIOS authors, "[e]ffective governance seeks to foster good science, not to hamper it, but recognizes that good science goes hand in hand with open, clear, transparent regulation to ensure both trust and accountability."[80] This nuanced approach should be considered and applied across each element of governance.

As synthetic biology moves out of the laboratory, several existing US regulations are likely to be applicable, including TSCA, the Community Right-To-Know Act and various air, water, solid waste, worker safety and consumer protection statutes.[81] TSCA already applies to the intentional release of genetically modified inter-generic microorganisms into the environment.[82] As with nanotechnology, research will be required to identify gaps or mismatches in existing regulation as it applies to synthetic biology. Due to the growth of the biotechnology industry and genetic modification in previous years, the EPA, USDA, and FDA already have a framework for regulating the use and commercial production of genetically modified microbes, plants, and food and drugs.[83] While not mandatory, the National Institutes of Health also have produced a set of guidelines for the research phase of recombinant DNA technology, which may be strengthened and more directly targeted toward current research.[84]

One potential downfall of our current regulatory framework, particularly in health and environmental protection, is its heavily science-based approach. TSCA, for instance, applies restrictions based the chemical identity of substances, but physical form and interactions of various chemicals can create unique hazards. Much like nanotechnology, the complex uses and uncertainties of synthetic biology may present a conundrum for the regulatory scheme. Sheila Jasanoff has described overly science-driven regulatory systems as creating "technologies of hubris," in which "the unknown, unspecified, and indeterminate aspects of scientific and technological development remain ... treated as beyond

---

[80]   Id. at 7.

[81]   See Synthetic Biology Project, supra note 19, at 27.

[82]   Toxic Substances Control Act, 15 US C. §§ 2601-2697 (2012).

[83]   For further discussion of the potential application of various regulations to synthetic biology, see Kenneth Oye, "The Regulation of Synthetic Biology: A Concise Guide to US Federal Guidelines, Rules and Regulations," SYNBERC 13–16 (2010), available at http://synberc.org/humanpractices/synbio-regulations-guide (accessed July 12, 2013).

[84]   National Institutes of Health, Guidelines for Research Involving Recombinant DNA Molecules (2002), available at http://oba.od.nih.gov/rdna/nih_guidelines_oba.html (accessed July 12, 2013).

reckoning, they escape the discipline of analysis."[85] By narrowly defining risks, such a system prevents contemplation of less visible risks.

Public engagement will be a key governance tool for synthetic biology, and public perception an important aspect of an integrated approach. In a recent representative poll in the US, however, 67 percent of the public had never heard the term "synthetic biology."[86] Research has also shown that communicating the risks of synthetic biology to the public may be a challenge; many have a more negative inherent reaction to synthetic biology than they do to nanotechnology. Interestingly, one study shows that segments of the population generally unconcerned by environmental risks such as climate change and nuclear power *do* show concern regarding synthetic biology.[87] Educational institutions and public interest organizations can play an important role in the process of communicating risks. One example of this is the Synthetic Biology Project at the Woodrow Wilson Center for International Scholars, which has created a web-based Synthetic Biology Scorecard that tracks governmental efforts to improve oversight of research and development. The "scorecard" monitors federal progress on the presidential commission's recommendations for synthetic biology governance as well as non-federal activity.[88] The synthetic biology industry has a fresh opportunity to take the lessons learned from nanotechnology and previous technologies like genetic engineering, to encourage upstream public engagement during the research phase. Such engagement:

> could influence research priorities, provide critical feedback on hypothetical future applications and, perhaps most importantly, be used to establish and test processes and mechanisms that will respond to or deal with issues as they arise. On this understanding, public engagement becomes a kind of governance.[89]

---

[85]   Sheila Jasanoff, "Technologies of Humility: Citizen Participation in Governing Science," 41 MINERVA 223, 223-4 (2003).

[86]   Organisation for Economic Co-operation and Development and The Royal Society, Symposium on Opportunities and Challenges in the Emerging Field of Synthetic Biology: Synthesis Report 39 (2010), available at http://www. allea.org/Content/ALLEA/Synthetic%20Bio/RS_NAS_OECD_SynBio%20 Symposium%20-%20Synthesis%20Report.pdf (accessed July 12, 2013).

[87]   See id. at 40.

[88]   Synthetic Biology Scorecard, supra note 70.

[89]   Synthetic Biology Project, supra note 19, at 22.

Upstream engagement could also have the essential function of "enhanc[ing] social resilience to scientific uncertainty."[90]

Industry-based groups have suggested mainly self-regulatory approaches, and while these are unlikely to sufficiently manage risks on their own they will be part of a successful integrative approach. Focusing on the area of synthetic genomics, which constitutes a key component of the synthetic biology movement, a Sloan Foundation-funded report by the J. Craig Venter Institute and others presented detailed policy options for different points of the industry. The main "intervention points" discussed were education and public engagement, training tools, and cooperative oversight bodies.[91] An experimentalist approach, where the national government sets benchmarks for these groups to meet but allows great flexibility in how they reach them with the goal of collecting data on possible routes for governance, may be an ideal compromise between self-regulation and a command-and-control approach. Indeed, the Venter Institute states that "[t]o keep pace with such a dynamic situation, policy makers may choose to adopt a framework of 'adaptive decision making'" allowing them to create a "… suite of options that match today's technologies, the magnitude of today's risks and benefits, and societal priorities."[92] The study also considered registration and licensing programs for DNA synthesizers, noting that in the context of biosecurity concerns such a program may be the most cost-effective method for governments to enhance security.[93]

As indicated by the Venter study, different categories of synthetic biology application may warrant different oversight regimes: there might not be an appropriate 'one size fits all' approach. This takes us back to the BIOS "art of governance" discussion, which emphasizes fluidity between the various realms of governance:

> Instead of tying the development of governance to the requirement of a perfect definition, an alternative view (and one from the perspective of the arts of government) may be to see the subject of governance not as an 'object', but as 'interactions'. The same research practice may be recognised as "synthetic biology" by some and as "chemistry" or "genetic research" by others. Instead of getting electronic engineers and geneticists to all concur on the labeling of certain types of activity, a more effective governance approach

---

90   Zhang et al., supra note 14, at 24.
91   See Michele S. Garfinkel et al., supra note 27, at 20.
92   Id. at 18.
93   See id. at 32–5.

may be to equip stakeholders with the techniques that will help enable effective coordination between those from very different and unfamiliar fields.[94]

A truly integrated governance approach will consider nanotechnology in tandem with synthetic biology, as both are emerging technologies that present common challenges. The IRGC sees nanotechnology, along with biotechnology and information technology, as areas for regulatory precedent for synthetic biology governance.[95] As described by BIOS,

> Every new technology opens up new situations. Yet this 'newness' is seldom completely novel or distinct. Modern technologies share many core characteristics and subsequently lead to many similar governance challenges. As has been noted by many social scientists and policy makers, it may be more beneficial to align seemingly separate social discussions to develop a more general approach addressing common themes. The governance of synthetic biology provides a good opportunity to develop such a general approach. As an emergent technology, synthetic biology presents a developing field in which new governance strategies can be explored and a comprehensive regulatory apparatus is yet to be established. As synthetic biology seeks to position itself at the convergence of multiple research disciplines, unsettled governance problems from contributing disciplines may become more acute while previously settled issues may need to be reinvestigated. Rather than chasing dozens of loose-ends from the onset, there is an essential need to reflect on the common themes of these dilemmas. In addition, the prevalence of conflicting, even contradictory, narratives for synthetic biology suggests that current evidence-based nation-state governmentality alone will not be sufficient to comprehend or address the reality of these developments. There is a need to adjust regulatory outlooks and seek complementary governance approaches.[96]

4.5 REGULATORY ACTIVITY

Nanotechnology is increasingly gaining the attention of regulators. Synthetic biology, however, has seen little regulatory activity to date. This section discusses some of the emerging regulatory approaches.

---

[94] Zhang et al., supra note 14, at 23.
[95] International Risk Governance Council, supra note 33, at 10.
[96] Zhang et al., supra note 14, at 28.

4.5.1   Government Regulation by the EPA

In 2009 the EPA completed its voluntary Nanoscale Materials Steward-ship Program. While it provided the agency with some useful information there remain significant health and safety data gaps.[97] In March of 2011 the White House Emerging Technologies Interagency Policy Coordin-ation Committee released a set of overarching principles to be used for regulation and oversight of emerging technologies. The report encour-aged the use of "rational, science- and risk-based regulatory approaches that would be based on the full array of a material's properties and their plausible risks and not simply on the basis of size alone."[98]

Although some questions have been raised about how well the EPA has treated the "full array" of nanotechnology, it has in the last two years taken a number of regulatory actions under TSCA. On April 4, 2012, the EPA promulgated significant new use rules (SNUR) under TSCA[99] for 17 chemical substances.[100] This included a SNUR for infused carbon nanostructures, which are generally used as an additive to create conduc-tive properties to reinforcements in composites. These rules became effective in June of 2012, requiring producers to gather information of their products relating to potential health effects. The use of SNURs to fill in information gaps for potentially high-risk nanomaterials shows a promising application of the current regulatory framework. In August of 2012, the EPA proposed another set of SNURs for seven perfluoroalkyl sulfonate (PFAS) chemical substances that have not yet entered the production and import phase.[101] It also proposed a SNUR for long-chain

---

[97]   US Environmental Protection Agency, Office of Pollution Prevention and Toxics, Nanoscale Materials Stewardship Program: Interim Report 28 (January 2009), available at http://epa.gov/oppt/nano/nmsp-interim-report-final.pdf (accessed July 12, 2013).

[98]   Office of Science and Technology Policy, RESPONSIBLE REALIZATION OF NANOTECHNOLOGY'S FULL POTENTIAL (2011), available at http://www.whitehouse.gov/blog/2011/06/09/responsible-realization-nanotechnologys-full-potential (accessed July 12, 2013).

[99]   The products met the concern criteria at 40 C.F.R. Section 721.170(b)(3)(ii).

[100]   See US Environmental Protection Agency, "Significant New Use Rules on Certain Chemical Substances," 77 Fed. Reg. 20,296, 20,308 (April 4, 2012).

[101]   Perfluoronated substances such as PFAS and PFOS are used for a variety of purposes, including textile cleaning, industrial cleaning, and as insecticides. See Bergeson and Campbell PC, "EPA Announces Proposed SNUR for Perfluoroalkyl Sulfonate and Long-Chain Perfluoroalkyl Carboxylate," REGULATORY DEVELOPMENTS BLOG (August 10, 2012), available at

perfluoroalkyl carboxylate (LCPFAC) chemical substances used for carpet treatments and in 2011 promulgated several others.[102] These developments indicate EPA's reliance on existing TSCA provisions to regulate chemical-related risks. The agency has also released guidelines for the testing of antimicrobial products, including product performance testing and ecological effects testing.[103]

The EPA has also made progress on using FIFRA to regulate nanomaterials in pesticide products. In 2011 it made available a notice announcing a proposed initial plan that would:

1. use section 6(a)(2) of FIFRA to obtain existing information regarding what nanoscale material is present in a registered pesticide product and its potential effects on humans or the environment; relying on a registrant to inform the agency of relevant information pertaining to their products;

2. obtain information on nanoscale materials in pesticide products using data call-in notices under FIFRA section 3(c)(2)(B), under which EPA sends a data call-in (DCI) request to a registrant requiring the registrant to provide additional data or other information, which the registrant may need to generate or compile, would also need to require the inclusion of this information with any application for registration of a pesticide product that contains a nanoscale material; and

3. create a new approach for how EPA will determine on a case-by-case basis whether a nanoscale active or inert ingredient is a "new" active or inert ingredient for purposes of FIFRA and the Pesticide Registration Improvement Act, even when an identical, non-nanoscale form of the nanoscale ingredient is already registered.[104]

---

http://www.lawbc.com/regulatory-developments:/entry/epa-announces-proposed-snur-for-perfluoroalkyl-sulfonate-and-long-chain-per/ (accessed July 12, 2013). This rule would amend 40 CFR § 721.9582 to add additional PFAS chemical substances to an already-regulated list. Id.

[102]   For updated news on nano-related technologies see Bergeson and Campbell PC, RECENT REGULATORY DEVELOPMENTS, available at http://www.lawbc.com/regulatory-developments (accessed July 12, 2013).

[103]   See US Environmental Protection Agency, Test Methods and Guidelines, available at http://www.epa.gov/ocspp/pubs/frs/home/testmeth.htm (accessed July 12, 2013).

[104]   US Environmental Protection Agency, REGULATING PESTICIDES THAT CONTAIN NANOMATERIALS, available at http://www.epa.gov/pesticides/regulating/nanotechnology.html (accessed July 12, 2013).

In December of 2011, EPA conditionally identified pesticide products containing nanosilver as new active ingredients under FIFRA.[105]

Also in 2012 the EPA, jointly with the USDA, Fish and Wildlife Service, and National Marine Fisheries Service, indicated that it intended to expand public engagement and multi-stakeholder collaboration in the regulatory review process. Focusing on the pesticide registration review process and endangered species consultations, they proposed:

- emphasis on coordination across these federal agencies;
- expanded role for USDA and the pesticide user community in providing current pesticide use information to inform and refine EPA's ecological risk assessments;
- "focus" meetings at the start of registration review for each pesticide active ingredient, to clarify current uses and label directions and consider the potential for early risk reduction;
- formal ESA consultations later in the registration review process, allowing time to engage stakeholders in the development of more refined ecological risk assessments and more focused consultation packages including mitigation for listed species;
- outreach to potentially affected pesticide users to discuss the technical and economic feasibility of draft Reasonable and Prudent Alternatives (RPAs), intended to avoid jeopardy to threatened and/or endangered species;
- descriptions of the processes by which (1) EPA will summarize and organize comments on RPAs and provide those comments to the Services and (2) the Services will prepare a document for the administrative record for the consultation explaining how comments were considered and, if appropriate, how the final biological opinion was modified to address the comments received.[106]

---

[105]   See US Environmental Protection Agency, Office of Pesticide Programs, DECISION DOCUMENT: CONDITIONAL REGISTRATION OF HEIQ AGS-20 AS A MATERIALS PRESERVATIVE IN TEXTILES (December 1, 2011), available at http://caat.jhsph.edu/programs/workshops/HeiQ%20Proposed%20Registration%20Decision%20Document%208-12-2010.pdf (accessed July 12, 2013).

[106]   PESTICIDE NEWS STORY, US EPA, http://www.epa.gov/oppfead1/cb/csb_page/updates/2012/espp-stakeholder.html (last visited February 17, 2013). See also PROPOSAL FOR ENHANCING STAKEHOLDER INPUT, U.S. EPA, available at http://www.epa.gov/oppfead1/endanger/2012/regreview-esa.pdf (accessed July 12, 2013).

4.5.2 State Action

Because progress toward a comprehensive federal approach to governing nanotechnology is moving slowly, some states have developed their own regulatory initiatives to govern nanotechnologies. California's Green Chemistry Initiative is designed to create a systematic, science-based process to evaluate chemicals of concern in consumer products and promote use of safer alternatives. California has played a leading role in gathering information on nanotechnology risks thanks to a 2006 amendment to the state Health and Safety Code authorizing the Department of Toxic Substances Control (DTSC) within the California Environmental Protection Agency and other agencies to request information "regarding analytical test methods, fate and transport in the environment, and other relevant information from manufacturers on chemicals of concern."[107] DTSC began gathering such information on nanotechnology in 2009 with a voluntary reporting initiative, when it "requested information regarding analytical test methods, fate and transport in the environment, and other relevant information from manufacturers of carbon nanotubes."[108] In 2010, DTSC required manufacturers to provide them with information regarding the chemical and physical properties, analytical test methods and other relevant information on several nanomaterials, including nano metals, metal oxides, and quantum dots.[109]

4.5.3 International Action in the EU, Canada, and Australia

The most applicable legislation for the regulation nanomaterials in the European Union is REACH (Registration, Evaluation, Authorisation and Restriction of Chemical Substances). One key difference between REACH and United States chemical regulations is that REACH uses a "no data, no market" principle and "requires every party in the user chain to generate and share information."[110] However, REACH currently lacks

---

[107] California Dept. of Toxic Substances Control, CHEMICAL INFORMATION CALL-IN OVERVIEW, available at http://www.dtsc.ca.gov/PollutionPrevention/Chemical_Call_In.cfm (accessed July 12, 2013).

[108] California Dept. of Toxic Substances Control, CARBON NANOTUBE INFORMATION CALL-IN, available at http://www.dtsc.ca.gov/PollutionPrevention/Round_One.cfm (accessed July 12, 2013).

[109] California Dept. of Toxic Substances Control, CHEMICAL INFORMATION CALL-IN – NANO METALS, NANO METAL OXIDES, AND QUANTUM D, available at http://www.dtsc.ca.gov/PollutionPrevention/Round_Two.cfm) (accessed July 12, 2013).

[110] Lee, *supra* note 10, at 809.

a specific mechanism to place requirements on registrants and notifiers of nanomaterials.[111] The European Chemicals Agency (ECHA) does have "a very large repository of REACH registration and C&L [Classification & Labeling] notification dossiers that could in principle provide information on nanomaterials registered or notified and therefore on the market," however a 2011 screening of dossiers brought up only 78 registered substances which contained some information on nanomaterials.[112] While the official position of the European Commission is that nanomaterials are covered by the REACH definition of "substance," there has also been discussion of treating them as a separate substance.[113]

Aside from traditional chemicals legislation, the EU has also developed some nanotechnology-specific requirements in certain areas that the US has not yet tackled. Article 16 of the EU Cosmetics Regulation now requires manufacturers of nano-containing cosmetics to notify the European Commission six months before placing them on the market. The Commission can also request an opinion from the Scientific Committee for Consumer Safety if ever "it has any concerns, for example due to new information supplied by a third party."[114]

Early in 2009, the Canadian government considered closing a regulatory gap that required reporting only for chemicals produced at above a one ton threshold. A mandatory survey proposed jointly by Environment Canada and Health Canada would have required institutions that manufacture or import more than one kilogram of nanomaterials to submit all of the information they have on the physical and chemical properties, toxicological data, and manufacturing methods and uses of the materials.[115] This requirement was not adopted, however. Canada currently regulates nanomaterials under its 1999 Environmental Protection Act,[116] Pest Control Products Act, Fertilizers Act, Feeds Act, and Food and

---

[111]   See European Chemicals Agency, Newsletter: REACHing 2013 16 (2012), available at http://echa.europa.eu/documents/10162/13584/echa_newsletter_2012_01_en.pdf (accessed July 12, 2013).

[112]   Id.

[113]   See id.

[114]   See Lee, supra note 10, at 814 (citing to Article 16(5) and 16(6)). Lee also notes that Article 16 does not apply to substances such as colorants, preservatives or UV-filters if they have already been positively listed on an Annex Id.

[115]   See Victoria Gill, "Nano-Regulation Creeps Closer," ROYAL SOCIETY OF CHEMISTRY (Feb. 25, 2009), available at http://www.rsc.org/chemistry world/News/2009/February/25020901.asp (accessed July 12, 2013).

[116]   Part 5 of the Act requires manufacturers of new substances to provide specific information to government officials so it can be evaluated for potential effects on human health and the environment. Health Canada, one of the agencies

Drugs Act. The federal government of Canada also funds domestic health and safety research pertaining to nanomaterials and "actively participat[es] in international efforts to study, quantify and understand the behavior and toxicity of nanomaterials."[117] The national framework was based on the principle that existing regulatory approaches and risk management strategies would suffice for nanotechnology so long as more work was done in the realm of strategic risk assessment research.[118]

Australia has taken significant strides in the regulation of nanotechnology. In 2008 the Australian Department of Innovation, Industry, Science and Research focused on the governance challenges imposed by nanotechnology, creating a National Nanotechnology Strategy and an Office of Nanotechnology.[119] In 2009 the Department replaced the Nanotechnology Strategy with a four-year plan that it calls the National Enabling Technology Strategy ("NETS").[120] Defining "enabling technologies" as "new technologies or new uses for existing technologies that enable new products or services or more efficient processes ..." to include nanotechnology and biotechnology, the Strategy sets broad goals for the responsible development of such technologies and gives authority to implement the Strategy to various sections within the Department.[121] One of these sections – the Enabling Technologies Public Awareness and Community Engagement ("PACE") Section – seeks to engage the public "via free public forums, information and education materials, and feedback mechanisms such as regular public attitude surveys and a free call information service ... ."[122] PACE has published several surveys about

responsible for review of products along with Environment Canada, assesses the type of hazard and level of exposure. Id.

[117] Government of Canada, NANOREGULATIONS, available at http://nanoportal.gc.ca/default.asp?lang=En&n=23410D1F-1 (accessed July 12, 2013).

[118] Id.

[119] See Australian Dept. of Innovation, SCIENCE, RESEARCH, AND TERTIARY EDUCATION, NATIONAL NANOTECHNOLOGY STRATEGY, available at http://www.innovation.gov.au/Industry/Nanotechnology/National EnablingTechnologiesStrategy/Pages:/NationalNanotechnologyStrategy.aspx.) (accessed July 12, 2013).

[120] See generally Australian Dept. of Innovation, SCIENCE, RESEARCH, AND TERTIARY EDUCATION, NATIONAL ENABLING TECHNOLOGIES STRATEGY (2009), available at http://www.innovation.gov.au/Industry/Nano technology/NationalEnablingTechnologiesStrategy/Documents/NETS_booklet. pdf (accessed July 12, 2013).

[121] Id. at 3–4.

[122] Australian Dept. of Innovation, SCIENCE, RESEARCH, AND TERTIARY EDUCATION, PUBLIC AWARENESS AND ENGAGEMENT, available

national attitudes towards nanotechnology and related risks and it recently conducted a survey on public perceptions of risk pertaining to use of sunscreens containing nanomaterials.[123] On the educational front, PACE also jointly maintains with the University of Melbourne an educational web resource entitled "TechNYou," established with a goal to "meet a growing community need for balanced and factual information on emerging technologies."[124] Such a program, if introduced in the United States, could have a significant impact in meeting goals of upstream public involvement.

4.6 CONCLUSION

Governance mechanisms have slowly developed for nanomaterials over the past few years with industry initiatives and some government regulation, especially regulation focused on information disclosure. Governance tools such as tort liability remain in place as a potential check on introducing new materials to the market without adequate precaution, despite some early efforts to limit tort liability for nanotechnology. However, no government has of yet put in place a comprehensive system of governance that includes anticipatory governance processes that are able to analyze the success or regulatory models and adjust as new information becomes available, consumer information disclosure, ongoing information disclosure to regulators, public information tools, and public dialogues.

In the context of synthetic biology even less progress has been made in developing an effective system of governance that allows experimentation and innovation while at the same time being able to identify and deal with risks to health and the environment. Perhaps the most important step that should be taken is developing through an open and transparent process a system of anticipatory governance that can deal with the risks of these new technologies. At the same time the industry must continue

---

at http://www.innovation.gov.au/Industry/Nanotechnology/PublicAwarenessand Engagement/Pages/default.aspx (accessed July 12, 2013).

[123] Australian Dept. of Innovation, SCIENCE, RESEARCH, AND TERTIARY EDUCATION, STUDY OF PUBLIC ATTITUDES TOWARDS SUNSCREENS WITH NANO PARTICLES (February 2012), available at http://www.innovation.gov.au:/Industry/Nanotechnology/PublicAwarenessand Engagement/Documents/SunscreenStudy.pdf (accessed July 12, 2013).

[124] See generally TechNYou, available at http://technyou.edu.au (accessed July 12, 2013).

efforts to self-regulate to protect both its reputation and the public. Finally, the emergence of synthetic biology makes finding new and better ways of engaging the public a critical part of the governance challenge.

5. The role of adaptation in the governance of emerging technologies

Marc A. Saner

5.1 THE CONTROL PARADIGM IN TECHNOLOGY GOVERNANCE

The context of this book – derived from workshops on the *pacing problem* and the *utility of soft law in governing emerging technologies* – provides a broad platform from which to consider the limits of technology governance. I argue that it is necessary to stretch the scope of the concept *governance* beyond its normal boundaries in order to fully address the challenges caused by rapidly emerging technologies. This perspective on governance entails that a focus on social adaptation (rather than on the control of technology) is necessary in many instances. I will also show that such focus on adaptation can – ironically – ultimately improve the control of emerging technologies. A few analogous examples from the adaptation to climate change will provide practical models or ideas for implementation.

'Governance' is a word with roots in the fourteenth century[1] although it has entered the public policy discourse only relatively recently, approximately 20 years ago. The concept is now often juxtaposed with "government" to emphasize that decision making power and accountability are frequently shared among governments, the private sector, non-government organizations, consumers, and the public at large. The following definition by the non-profit Institute on Governance expresses a succinct concept used by practitioners: "Governance determines who

---

[1] Merriam-Webster Dictionary, "Governance", available at http://www.merriam-webster.com/dictionary/governance (accessed October 20, 2012).

has power, who makes decisions, how other players make their voice heard and how account is rendered."[2]

An important governance element that is perhaps too obvious to be included in a definition is that one always governs toward a goal. In other words, the idea of performance is taken as a given; everyone knows that the governing parties have goals that will define how power is used, how decisions are shaped, how outside voices are heard and used, and how accountability standards are designed and implemented. In sum, it's a given that risk and opportunity (pain and gain) will drive how governance models are designed and how existing governance models are used.

An important aside: the focus on *control and accountability* in public administration often comes in conflict with *performance*. The potential conflict needs to be emphasized because the tools of choice to foster accountability are results based management (RBM) and performance measurement. It appears, thus, that a focus on control and accountability goes hand-in-hand with a focus on performance measurement. However, a strong implementation of performance measurement and other account-ability tools can have two undesired effects. First, it can increase the bureaucratic load through an increase in reporting and the need for indicator development and deployment. Second, it can increase risk adversity through a stronger awareness of being watched and a replace-ment of the value of personal judgment with whatever incentives are fostered by the performance indicators. In brief, a public service that is busy with measurement and fearful of accountability issues can easily be less productive despite the apparent focus on performance.

The explicit and "enlightened" consideration of performance, however, is important. Without it, one could easily view "governance" as a term that is entirely about process, a system of control, or even a regulation. Even adaptive regulation and governance[3] could be understood as a control mechanism – albeit one that is particularly dynamic and, there-fore, more compatible with rapidly changing environments. But what can we do when we reach the limit of control? For example, how should governance be conceived in those instances where technological diffusion renders international rule setting and enforcement virtually impossible (think about the difficulties the *war on drugs* is facing despite its strong

---

[2]   Institute on Governance, DEFINING GOVERNANCE (undated), available at http://iog.ca/en/about-us/governance/governance-definition (accessed October 20, 2012).

[3]   Gary E. Marchant, Douglas J. Sylvester and Kenneth W. Abbott, "What Does the History of Technology Regulation Teach Us about Nano Oversight?," 37 J. LAW MED. ETHICS 724, 724–26 (2009).

commitment to prohibition)? Or, how can control mechanisms be designed when the environment changes faster than even well-funded control agencies can adapt?

I argue that some instances of emerging technologies provide such challenges. For example, the declining cost of DNA synthesis[4] is driving a technological diffusion that has been called "biohacking" by *The Economist*.[5] It does not require further explanation to understand that as biotechnologies become more modular and cheap, they will become more diffused internationally and that such diffusion will make the implementation of control mechanisms more difficult if not impossible. The same forces that enable diffusion often also drive technological acceleration. There are several technologies that are outpacing control mechanisms. Nanotechnologies are a well-studied example.[6]

In addition to the spatial issue (diffusion) and the temporal issue (pacing), one could add a third kind: those technological developments that raise foundational ethical issues. By "foundational" I mean those issues that require a metaphysical debate beyond concerns over potential inefficiencies and inequalities. An example is the move from human modification as it is practiced today (say, plastic surgery, prosthetics and the consumption of mild stimulants such as coffee) to transhumanism (say, cyborg style implants, extreme prosthetics, genetic engineering, and the consumption of truly performance altering substances). One could even label these developments "game changers." The radical transformation of ourselves, the agents in the governance exercise, renders the entire debate teleologically open – the alteration of the agents is akin to changing the cause during the assessment of a cause–effect.[7]

The control paradigm, thus, has its limits when diffusion, pacing and ethical issues associated with emerging technologies become significant,

---

[4]   Robert Carlson, "The Changing Economics of DNA Synthesis," 27 NATURE BIOTECHNOLOGY 1091, 1091 (2009).

[5]   "Biohacking: Hacking Goes Squishy," THE ECONOMIST TECH. QUART. Q3 (September 3, 2009), available at http://www.economist.com/node/14299634 (accessed October 20, 2012).

[6]   Gary E. Marchant, "The Growing Gap Between Emerging Technologies and the Law," in THE GROWING GAP BETWEEN EMERGING TECHNOLOGY AND LEGAL-ETHICAL OVERSIGHT: THE PACING PROBLEM 19, 25–6 (Gary E. Marchant, Braden R. Allenby and Joseph R. Herkert, eds, 2011).

[7]   For a discussion of the "technoself," see Marc A. Saner and Jeremy Geelen, "Identity in a Technological Society: Governance Implications," in HANDBOOK OF RESEARCH ON TECHNOSELF: IDENTITY IN A TECHNOLOGICAL SOCIETY 720, 720–41 (Rocci Luppicini, ed., 2012).

as it is often the case. This view is in line with McCray and Oye who state: "our core argument for emphasizing adaptation is based on pessimism with respect to anticipation."[8] The principal danger is loss of time – a significant cost in a context where pressure mounts as time passes. It builds in the way shown in Table 5.1, below.

Table 5.1   *Non-routine pressures and the six steps of how the technology control paradigm can fail – and likely will fail in some emerging technology contexts*

Step	Step-wise control response
1	A response in the technology governance system is triggered due to non-routine pressures (high rates of diffusion/pacing or significant ethical issues).
2	Standard-type control mechanisms (for example, versions of existing regulations) are attempted.
3	Standard-type control mechanisms fail at the stage of creation or implementation because they are recognized as insufficient.
4	Alternative-type control mechanisms (for example, soft governance mechanisms such as self-regulation) are attempted.
5	Alternative-type control mechanisms fail at the state of creation or implementation; they are too weak, not suitable for the magnitude of the pressure.
6	Debates and research keep governance experts in business while providing communication content for those who have to soothe dissatisfied technology watchers. The incentives inside the system to abandon the control paradigm are weak and the process is prolonged.

In this chapter, I argue that it is in our collective interest to understand the magnitude of the pressure early on so that we can select our approach correspondingly. For small issues, the procedure above will be successfully resolved after step (2) or after step (4) at the latest. For large issues,

---

[8]   Lawrence McCray and Kenneth A. Oye, ADAPTATION AND ANTICIPATION: LEARNING FROM POLICY EXPERIENCE 3, paper presented at the NSF-EPA Transatlantic Uncertainty Colloquium, Washington DC (June 29, 2007), available at http://kms1.isn.ethz.ch/serviceengine/Files/ISN/93715/i publicationdocument_singledocument/f5345751-3f44-4555-8541-a1bc03bacb11/ en/mccrayoye-petpworking.pdf (accessed October 20, 2012).

the procedure will not succeed and the earlier we abandon it, the better because time is of the essence and delays build additional pressure.

5.2   A COPERNICAN REVOLUTION IN TECHNOLOGY GOVERNANCE

It is worth noting at this point that several scholars view technology completely outside of a "governing technology" paradigm. Instead, they emphasize a "being governed *by* technology" paradigm. I use the word paradigm here because the switch is rather akin to the switch from Ptolemaic to a Copernican worldview. We may believe that "Planet Technology" will naturally circle its creator and central star, which we could call "Human Spirit." This quasi-Ptolemaic view can be replaced by a quasi-Copernican view in which a relatively small "Planet Human" appears to be circling a vast "Technology Star."

This idea is already present in Langdon Winner's work on autonomous technology[9] and his work on the question if artifacts have politics.[10] More recently, Susan Blackmore has coined the term *temes* to label technological units of evolution that act in an apparently selfish fashion, in analogy to Richard Dawkins' concept of *memes* that are themselves modeled after so-called "selfish *genes*."[11] Kevin Kelly, former editor of *Wired Magazine* makes a similar point when he states: "the technium [the technological system around us] has its own wants."[12] Both authors, Blackmore and Kelly, bring holistic, evolutionary thinking to the discussion.

Evolutionary thinking, that is, thinking about the survival of the fittest, naturally leads to a strong consideration of performance (utility) over control (rules) despite the fact that modern Darwinists like Susan Blackmore or Daniel Dennett[13] will spend a lot of effort illustrating how *design without designer* (a form of apparent control) can emerge simply

---

[9]   Langdon Winner, AUTONOMOUS TECHNOLOGY: TECHNICS-OUT-OF-CONTROL AS A THEME IN POLITICAL THOUGHT (1977).

[10]   Langdon Winner, THE WHALE AND THE REACTOR: A SEARCH FOR LIMITS IN AN AGE OF HIGH TECHNOLOGY (1986).

[11]   Susan Blackmore, "Dangerous Memes; or, What the Pandorans Let Loose," in COSMOS & CULTURE: CULTURAL EVOLUTION IN A COSMIC CONTEXT 297, 305 (Steven J. Dick and Mark L. Lupisella, eds, 2010).

[12]   Kevin Kelly, WHAT TECHNOLOGY WANTS 15 (2010).

[13]   Daniel C. Dennett, DARWIN'S DANGEROUS IDEA: EVOLUTION AND THE MEANINGS OF LIFE (1995).

by combining heritability, variation, selection and iteration. The possibility that technology has its own wants, or that it may be a unit of evolutionary selection that is quite autonomous, should be taken seriously. It adds further argument to the belief that the control paradigm needs to be enhanced or replaced in the context of the governance of emerging technologies.

5.2.1 Finding the Right Perspective

Technology governance is neither black nor white. While it is helpful to recognize that to some extent we *are* governed by technology, we also need to recognize that we *do* successfully control many aspects of the technological system. Standard and alternative-type control mechanisms (Table 5.1), thus, are productive in some instances but we also need the ability to understand when to do what, to recognize failure, and to progress successfully after failure.

It will be helpful at this stage to tabulate the common stances one may take in response to pressures caused by technological processes or by the products of technology – be those pressures physical, administrative, economic, moral, domestic political, or international. The nine stances shown in Table 5.2 provide a rather rich menu of options that go far beyond a narrow concept of control.

The menu of options in Table 5.2 cannot summarize the complex realities of governance processes. Mixtures of stances may be taken. Single individuals or even isolated teams will hardly ever make important decisions regarding technology governance. The knowledge base and policies can change rapidly. It's unavoidable, thus, that the menu of stances is more descriptive of the components of on-going debate rather than say, a national strategy. The purpose of this menu is to clearly show that many stances are reasonable and that there is a place for the final stance in this table: *adaptation*. Here follows a brief description of the space that avoidance, control, and adaption hold in technology governance.

5.2.2 Avoidance

These stances can be very meaningful early in technological development. A new technological platform may be named in the innovation context (say, "nanotechnology" or "synthetic biology") but it is not in the least clear (a) that new pressures, such as novel risks and novel ethical issues, are emerging, (b) to what extent there is equivalence in the

Table 5.2   Common types of stances taken in response to technology pressures

Stance	Characterization	Paradigm
Deny	"There is no pressure" – or: "The benefits and risk are equal (as far as we can tell)"	*Avoidance*
Avoid	"Not our problem; by the time it really hits, we will be elsewhere and everything will be different"	
Tolerate	"We cannot deny this pressure, but we have other priorities first; we will watch it but not interfere (yet)"	
Paralysis (I)	"Ok, it is important and urgent – but it's not feasible to attempt *anything* – we lack the power or tools (or even the motivation)"	
Manage	"It's important, urgent and action is feasible; we should modify, lessen or silence the pressure when it comes to us"	*Control*
Delegate	"It's important, urgent and action is feasible, but someone else should deal with it"	
Resist	"It's important, urgent and action is feasible; we should modify, lessen or silence the pressure *at the root*" [= heroic moment]	
Paralysis (II)	"Nothing is working but we just need to research, innovate and debate more; we know that our ingenuity is all we have now"	
Adapt	"It's important and urgent but we are not able to significantly lessen or silence the pressure; let us change the way we receive the pressure and how we live with it" [= acceptance moment]	*Adaptation*

Note: ordered as a progression or increasing acceptance of the combined reality of several factors: reality of the pressure, importance and magnitude, urgency, and difficulty of coping with the pressure. "Paralysis I and II" are non-deliberate and include paralysis by analysis.

anticipated benefits and risk, or (c) to what extent the so-named platform is the right unit of discussion in the technology governance discourse.

(a) Instinctively one may think that if an innovation has the power to disrupt an industry, then it should also have the power to endanger human health or the environment. It is indeed odd to hear the CEO of a company argue: "this technology will change *everything*" and then hear the same individual say "but we do *not* need any new forms of regulatory oversight." Yet, there are technologies, for example the transistor, which are simultaneously disruptive to industry and neutral (or even beneficial) to human health and the environment.

(b) Sometimes the balance of benefit and cost is truly symmetrical due to utter ignorance. We may thus find it difficult to decide if we ought to promote or stifle technology. In those cases where the potential to endanger human health and the environment are plausible, risk-relevant research may be commissioned. Such early research into human and environmental risk will appear more as subsidies to the innovation system than as attempts to control the technology. Stances such as "deny or tolerate" may therefore denote the denial of the need for regulatory action, but not the denial of the need for planning and related actions.

(c) When a new technology emerges it will normally first be framed in the context of innovation, for example as part of a new government program to subsidize research in a particular field. Such framing may not translate well into the regulatory context. Many elements of a new platform can be promising for inventors but only a few deserve new regulatory attention. This can lead to an avoidance stance because the demand for a governance discussion is not convincing to experts when watchdog groups target the entire (innovation) platform rather than (regulatory) problem areas.

The paralysis step within the avoidance stances is perhaps the hardest to defend. It does have its place when budgets are extremely tight as it may be the case in many developing nations. The wait and see approach has merit when there is a real chance that a "free-rider" approach is the best available compromise. To illustrate this point, regulators in developing nations may argue that an approval by the Food and Drug Administration of the United States can serve as a proxy and that it would be very expensive to best the quality of the assessment. While it is possible to defend an avoidance stance in some circumstances, the associated risk is a likely deterioration of public trust. This risk can be very significant and it may be contagious in the sense that skepticism of how scientists "monkey around in their labs" can become generic, independent of technology platforms and range across the political spectrum.

5.2.3   Control

In some instances, those engaged in the technology governance dialogue will not even consider any version of avoidance because the three stances listed under "control" in Table 5.2 provide a rich environment for strategy and action or because they are mandated to regulate. Adaptive forms of regulatory management and governance can use any of these three stances contextually, which is a significant advantage in situations where the knowledge base changes rapidly. They provide opportunities for adaptive and enabling regulations (especially under "manage"), for experimentation with novel approaches such as soft law (especially under "delegate"), and for calls for moratoria or a rethinking of the governance system (especially under "resist"). Models can be centralized or decentralized, command-and-control or collaborative, bureaucratic or flexible, secretive or transparent, legalistic or not.

However, the richness of the toolkit within the control paradigm does not guarantee that a governance system will successfully cope with all challenges. Instead, emerging technologies may overwhelm the system by speed (pacing), by distribution (diffusion), or by the novelty of ethical challenges, as discussed above. Although some of the proposed measures may feel truly heroic because of the way they challenge the *status quo*, the paralysis step can easily emerge as a result of control stances. The legitimacy of the control paradigm closes the mind to additional options, while the magnitude of the challenges given by some emerging technologies renders the chance of quick success very low. As indicated in Table 5.1, debates and research keep governance experts in business while providing communication content for those who have to soothe dissatisfied technology watchers. The incentives inside the system to abandon the control paradigm are weak and the process is prolonged.

The associated risk of ineffective action, experimentation and slow results lies in the mounting pressure over time. Mounting pressure may magnify both the challenges and public expectations. It can lead to political backlash. This, in turn, may magnify the risk of forced, poor or purely politically-driven action, the risk of real or perceived failure, and the deterioration of public trust. As a result, other stances and approaches may be needed.

5.2.4 Adaptation

The argument for adaptation follows McCray and Oye's[14] pessimism with respect to anticipation which entails skepticism regarding the utility of *a priori* rules and, thus, provides strong reasons to consider adaptive planning, regulation and governance. My argument here goes farther in several ways. I will make the case for adaptation by providing a set of 10 reasons:

1. Pessimism with respect to technological foresight not only provides an argument for adaptive regulation (which is still within the control paradigm) but also for acceptance that *social adaptation at a grand scale* may be unavoidable.

2. Regulation is often narrowly focused on risk control rather than the maximization of *benefits*. The planning for social adaption can be more encompassing than the narrow and legalistic framework of regulation.

3. The practice of public policy almost always targets *incremental* rather than transformational change. Public policy making in practice is therefore inherently weak when it comes to managing sudden social change.

4. Adaptation is a helpful backstop, even when the ability to anticipate the effects of an emerging technology appears sound. Rules may be carefully drafted based on sound anticipation research, but *implementation may be poor.*

5. The logic of *precaution* applies to the context of major emerging technologies because the *stakes are very high* (risk of lost opportunity, irreversible or significant risks to human health or the environment, damage to public trust and to the viability of administrative traditions), and cost-effective measures may be found in the domain of social adaptation.

6. The previous reason could also be expressed as an attitude of *humility*.[15] Also, if the "technium" and "temes" govern us more than we thought, then we should match this reality with the adaptation stance.

7. The "resist" stance, to rethink governance models from scratch or to consider moratoria is quite common in the context of emerging

---

[14] McCray and Oye, supra note 8.
[15] Sheila Jasanoff, "Technologies of Humility: Citizen Participation in Governing Science," 41 MINERVA 223, 223-44 (2003); Sheila Jasanoff, "Technologies of Humility," 450 NATURE 33, 33 (2007).

technologies, which leads to *polarization* and paralysis in the ability to debate. Serious efforts on adaptation will address governance needs during paralysis.

8. Adaptation is not just the final step when nothing else works; it can also represent the *cure against avoidance,* because it offers an alternative to the control paradigm and to mitigation, which may appear intractable.

9. Serious work on adaptation will ultimately (and ironically) *improve the quality of control* because once the limits of "manage, delegate or resist" are accepted then these approaches can be better planned. For example, a need for capacity building may become more obvious. Holistic planning has a greater chance to be realistic, systematic and calm. The ebb and flow or paralysis and frantic action can be avoided. Policy *experimentation* and learning are a form of adaptation that requires consideration beyond the regulatory mindset.

10. Adaptation is a useful addition, not a replacement, for the other stances in Table 5.2. It is perhaps the best stance to foster *holistic thinking* within this menu. Adaptive thinking can be applicable to all other stances by fostering a healthy criticism in at least two domains, strategy and capacity. For example, adaptation thinking may reveal that innovation and regulatory strategies are insufficiently integrated (and that benefits and risks cannot be properly balanced because administrative structures have not adapted to new and perhaps convergent technologies). Adaptive thinking is also eminently compatible with a focus on human resource strategies and capacity building.

If one accepts at least some of these reasons then one will want to know how technological adaptation could be approached. In practice, climate change is a theme where much emphasis has been put into debate on mitigation vs. adaptation. The debate over technology governance and climate change shares a few similarities. If one replaces the word "pressures" and "it" with "climate change" in Table 5.2 – the same pattern of stances holds. Stances are varied and confrontational. Data is lacking and forecasts are controversial. Climate change is as unavoidable as social change resulting from *temes,* the *technium,* or even a single invention such as *Facebook.* Even if the analogy is not perfect, it ought to be of interest to study current climate change adaptation measures and reflect on their utility in the technology governance context.

5.2.5 Learning from the Climate Change Analogy

The case of climate change provides nearly half a century of experience with a complex debate over the importance of the pressure for, and approaches to, mitigation and adaptation. As a result, the literature is vast. I selected here five examples that appear to be particularly interesting with an eye to porting ideas to the technology governance context.

5.2.5.1 Climate change adaptation research

The United Nations Framework Convention on Climate Change (UNCCC)[16] and, especially, the Intergovernmental Panel on Climate Change (IPCC), carry out research on climate change adaptation. The IPCC was established by the United Nations Environment Programme (UNEP) and the World Meteorological Organization (WMO) in 1988 and provides "policy-relevant and yet policy-neutral, never policy-prescriptive" analysis and recommendations.[17]

A recent key IPCC paper by Klein at al.[18] addresses the interplay between mitigation and adaptation. The paper identifies four types of inter-relationships between adaptation and mitigation: (1) adaptation actions that have consequences for mitigation; (2) mitigation actions that have consequences for adaptation; (3) decisions that include trade-offs or synergies between adaptation and mitigation; and (4) processes that have consequences for both adaptation and mitigation. The growing literature on this topic could provide inspiration and perhaps even evidence for the technology governance context.

---

[16]   The UNCCC provides a useful factsheet on adaption at http://unfccc.int/files/press/application/pdf/adaptation_fact_sheet.pdf (accessed October 20, 2012).

[17]   IPPC, ORGANIZATION (undated), available at http://www.ipcc.ch/organization/organization.shtml#.UM6XSaU48dI (accessed October 20, 2012).

[18]   R.J.T. Klein et al., "Inter-relationships between Adaptation and Mitigation," in CLIMATE CHANGE 2007: IMPACTS, ADAPTATION AND VULNERABILITY. CONTRIBUTION OF WORKING GROUP II TO THE FOURTH ASSESSMENT REPORT OF THE INTERGOVERNMENTAL PANEL ON CLIMATE CHANGE 745, 745–77 (M.L. Parry, et al., eds, 2007), available at http://www.ipcc.ch/pdf/assessment-report/ar4/wg2/ar4-wg2-chapter18.pdf (accessed October 20, 2012).

5.2.5.2 Climate change adaptation framework

At the United Nations Climate Change Conference in Cancun, Mexico, in 2010, parties adopted the *Cancun Adaptation Framework*[19] and affirmed that adaptation must be addressed with the same level of priority as mitigation. The framework includes five clusters of themes: (1) implementation – for example, the formulation and implementation of "national adaptation programmes of action (NAPAs)"; (2) support – such as financial, technological and capacity building support to developing countries, in particular those that are most vulnerable; (3) institutions – an Adaptation Committee at the global level (see below for details), regional centers and networks, and national-level institutional arrangements; (4) principles – for example, country-driven, gender-sensitive, participatory, transparent, science-based, respectful of traditional and indigenous knowledge, taking into consideration vulnerable groups, communities and ecosystems, with a view to integrating adaptation into relevant social, economic and environmental policies and actions; and (5) stakeholder engagement – relevant multilateral, international, regional and national organizations, the public and private sectors, civil society and other relevant stakeholders are invited to undertake and support enhanced action on adaptation at all levels.

This framework is the result of a global dialogue. Climate change is, of course, a global issue. One can argue that some aspects of emerging technologies are equally global with respect to benefits and risks, and with respect to the impact on economic, social and natural capital.

5.2.5.3 Climate change adaptation committee

As part of the Cancun Adaptation Framework (see above), a global Adaptation Committee was established to "promote the implementation of enhanced action on adaptation in a coherent manner" by providing technical support and guidance, by sharing relevant information, knowledge, experience and good practices, by promoting synergy and strengthening engagement with national, regional and international structures and by providing information and recommendations, drawing on adaptation good practices, for consideration by the international community.[20]

---

[19]   United Nations Framework Convention on Climate Change, CANCUN FRAMEWORK CONVENTION (undated), available at http://unfccc.int/ adaptation/cancun_adaptation_framework/items/5852.php (accessed October 20, 2012).

[20]   United Nations Framework Convention on Climate Change, ADAPTATION COMMITTEE (undated), available at http://unfccc.int/adaptation/

A formal structure such as that provided by the Adaptation Committee enables the effective sharing of information such as a database on local coping strategies.[21] There is no reason why an analogous exchange of information on social adaptation would not be of benefit in some emerging technology contexts.

5.2.5.4 Climate change adaptation fund

The UNCCC Adaptation Fund[22] finances concrete adaptation projects and programs in developing, vulnerable countries participating in the Kyoto Protocol. The fund is supervised and managed by the Adaptation Fund Board, and the World Bank serves as trustee. The fund provided support to over 30 nations in the amount of $200 million (total) during the period of April 2010 to November 2011.[23]

If social adaptation to emerging technologies would more explicitly enter the national and international governance and policy discourse, then the importance and full implications of the technology transfer component of international development would be recognized and, hopefully, coordinated.

5.2.5.5 Adaptation private sector initiative

The UNFCCC also started an Adaptation Private Sector Initiative (PSI), which "aims to catalyze the involvement of the private sector in the wider adaptation community."[24] The aim is to achieve a multi-sectoral partnership among governmental, private and non-governmental actors by providing "a platform for businesses to contribute in a sustainable and profitable manner to a strong and effective response, both in their own

cancun_adaptation_framework/adaptation_committee/items/6053.php (accessed October 20, 2012).

[21] United Nations Framework Convention on Climate Change, DATABASE ON LOCAL COPING STRATEGIES (undated), available at http://maindb. unfccc.int/public/adaptation/ (accessed October 20, 2012).

[22] United Nations Framework Convention on Climate Change , ADAPTATION FUND (undated), available at http://unfccc.int/cooperation_and_support/financial_mechanism/adaptation_fund/items/3659.php (accessed October 20, 2012).

[23] United Nations Framework Convention on Climate Change, REPORT OF THE ADAPTATION FUND BOARD, APPENDIX II (2011), available at http://unfccc.int/resource/docs/2011/cmp7/eng/06a01.pdf (accessed October 20, 2012).

[24] United Nations Framework Convention on Climate Change, ADAPTATION PRIVATE SECTOR INITIATIVE (undated), available at http://unfccc.int/adaptation/nairobi_work_programme/private_sector_initiative/items/4623.php (accessed October 20, 2012).

adaptation efforts and, importantly, in those of the most vulnerable countries and communities around the world."[25] The initiative has currently over 260 partners of which over 60 are classified as "private/ business sector," including major corporations such as Nestle, Pepsi, Microsoft, Google, BASF, and Allianz, to name just a few.[26]

Considering that the private sector is a major stakeholder in technology development, this particular initiative may be less portable to the emerging technology context. It is included here to show that once adaptation thinking is taken seriously, it can go in many directions.

5.3   CONCLUSION

This chapter shows that the control paradigm is too limited to address all issues that arise in the context of emerging technologies. Instead, an entire host of perspectives needs to be considered, and social adaptation to technological change is an important perspective when temporal scales (pacing), spatial scales (diffusion), or novel ethical issues put extraordinary pressures on the existing governance system.

The analogy of climate change adaptation shows that many practical projects can be pursued within an adaptation framework. Some of these could be studied in detail and proposals for the technology governance context could be developed on that basis. To identify just one example, in analogy to the Climate Change Adaptation Fund and to the ELSI programs established in the biotechnology contexts (at least in the US and in Canada), why not tie-in grants in the natural sciences with an ELSI component that explicitly covers adaptation research? Anything that would balance the debate away from a "mitigation-only" focus should be welcome. Anything that emphasis the global nature of the issue should be welcome too.

Ultimately, the most tangible benefit from the explicit attention to social adaptation for emerging technologies may lie in improvements to capacity building and engagement. Examples are upstream engagement of scientists and downstream engagement of end-users, improvements to science literacy and improvements to policy literacy, greater maturity

---

[25]   Id.

[26]   United Nations Framework Convention on Climate Change, NAIROBI WORK PROGRAMME (NWP): PARTNERS, ACTION PLEDGES AND EXPERTS DATABASE (undated), available at http://unfccc.int/adaptation/ nairobi_work_programme/partners_and_action_pledges/items/5005.php (accessed October 20, 2012).

when it comes to balancing local risks with global benefits, better planning for regulatory capacity, education, and improved communication at the science/policy interface.

Considering the complexity of the technology governance file – and its importance both in terms of benefits and risks – it should be desirable to improve these capacities and engagement tools. Good governance requires such foundations for the full range of governance tools, including adaptation.

5.4 ACKNOWLEDGMENTS

The author acknowledges helpful insights from Tim Plumptre (biotech governance), Jackie Dawson (climate change adaptation), Gary Marchant (invitation and discussions) and Lorena Ziraldo (comments and discussions).

6. Integrating technology assessment into government technology policy

Timothy F. Malloy

6.1 INTRODUCTION

There is little question that we face substantial challenges in the twenty-first century: climate change, disease, poverty, natural resource depletion and degradation, war and terrorism – the list goes on. Emerging technologies will likely be integral in resolving or mitigating many of these challenges; prominent examples include renewable energy technologies, green chemistry, biotechnology, nanotechnology, and informatics. Government policies often play a significant or even central role in advancing the development and diffusion of such technologies. Yet these policies typically fail to account for the unintended adverse health, environmental, social and other consequences that may flow from those technologies. This chapter starts with the premise that the United States' technology policy ought to integrate principles of protection and promotion so as to ensure the availability of truly beneficial technologies. It examines the extent to which this integration can be accomplished through technology assessment at the legislative stage of policy formulation. For these purposes, technology assessment refers to the systematic assessment and evaluation of the positive and negative impacts of an ostensibly beneficial technology. Technology assessment has a decades-long history, starting primarily in the United States and more recently establishing itself in Europe. It takes many forms, not all of which will be appropriate for use in legislative settings in the United States.

After some definitional matters regarding the notion of technological change, the chapter turns to two specific aspects of government technology policy. Promotional policies actively support the development and use of emerging technologies through research funding, direct subsidies and other strategies. Regulatory policies create a demand for emerging technologies by removing market competitors, creating a need for the emerging technology, or even requiring its use. Next, the chapter examines the problem of unintended consequences by identifying potential causes and exploring the historical example of methyl tertiary-butyl ether (MTBE) and the Clean Air Act. The chapter then briefly surveys the field of technology assessment, and considers how it could mitigate the problem of unintended consequences in the legislative setting. It concludes that, given the institutional structure and dynamics of Congress, technology assessment can address some, but not all, causes of such unintended consequences.

6.2   TECHNOLOGICAL CHANGE AND TECHNOLOGY POLICY

The concepts of technological change and innovation are widely discussed, but neither has a single definition.[1] Many commentators define technological change as a three-step process consisting of "invention" (the idea or model for a new process or device), "innovation" (the initial commercial application of the invention) and "diffusion" (the spread of the innovation across the relevant market sector).[2] This chapter adopts the more articulated, six-step concept of the innovation–development process described by Rogers in his classic *Diffusion of Innovations*.[3] The innovation–development process begins with the identification of needs or problems, followed by the next stage of basic and applied research. Basic research involves original investigations intended to advance scientific knowledge generally, while applied research is directed at meeting particular needs. Development follows closely behind research, taking the idea generated from basic and applied research and putting it into a form – such as a prototype device or a bench scale process – that is expected to meet the needs of the ultimate potential end-users.[4] Commercialization is the production and marketing of the newly developed product or process. It in turn is followed by diffusion – the process by which

---

[1]   Timothy F. Malloy, "Regulating by Incentives: Myths, Models and Micromarkets," 80 TEX. L. REV. 532, 539–40 (2002–02); David Wallace, ENVIRONMENTAL POLICY AND INDUSTRIAL INNOVATION 11 (1995).

[2]   Adam B. Jaffee and Robert N. Stavins, "Dynamic Incentive of Environmental Regulations: The Effect of Alternative Policy Instruments on Technology Diffusion," 29 J. ENVTL. ECON. & MGMT S-43, S-45 (1995); Paul Stoneman, THE ECONOMIC ANALYSIS OF TECHNOLOGICAL CHANGE 8 (1983).

[3]   Everett M. Rogers, DIFFUSION OF INNOVATIONS 132–50 (1995).

[4]   Id.; Benoît Godin, "The Linear Model of Innovation: The Historical Construction of an Analytical Framework," 31 SCI. TECH. & HUMAN VALUES 639, 647 (2006).

knowledge about the product or process is communicated to members of
the relevant sector – which, if successful, would lead to broad adoption.[5]
The final phase of the innovation–development process reveals the
consequences of technological change, with respect to both its impact
upon the original problem, and its other beneficial or adverse effects,
such as social, economic, health, or environmental impacts.[6]

Rogers' outline of the innovation–development process does not reflect
the diversity, complexity and nuance of technological change in the real
world. Nonetheless, the innovation–development process outline remains
useful as a conceptual tool, particularly here in analyzing the different
ways in which government technology policy can affect technological
change. To that question we now turn. The federal government plays a
substantial role in technology advancement through a variety of programs. In many cases, the goals of such programs are ensuring economic
growth and securing the nation's place globally as a leader in emerging
technologies.[7] In other cases, the funding focuses upon encouraging
technological advances needed to advance an agency-specific mission, or
other non-economic goal of the federal government. For example, the
Environmental Protection Agency (EPA) funds research and development
efforts for new remediation techniques needed to implement the Super-
fund program. Often, both types of goals are cited together by government officials when justifying federal expenditures on research and
development.[8] In practice, the federal government advances technology
change in two major ways. First, the government engages in "technology-
push" efforts; that is, actively supporting the innovation–development
process through direct or indirect funding.[9] Direct funding typically takes

---

[5]   Rogers, supra note 3, at 5–6, 143.

[6]   Id. at 150.

[7]   Godin, supra note 4, at 643–4; Matt Hourihan, "Federal R&D in the FY
2013 Budget: An Introduction," in AAAS REPORT XXXVII: RESEARCH AND
DEVELOPMENT FY 2013 5, 11–14 (2012).

[8]   See, for example, White House Office Of Science And Technology
Policy, INNOVATION FOR AMERICA'S ECONOMY, AMERICA'S ENERGY,
AND AMERICAN SKILLS: SCIENCE, TECHNOLOGY, INNOVATION, AND
STEM EDUCATION IN THE 2013 BUDGET (February 13, 2012), available at
http://www.whitehouse.gov/sites/default/files/microsites/ostp/fy2013rd_summary.
pdf (accessed July 16, 2013) (justifying increased R&D funding for a variety of
federal agencies).

[9]   Gregory F. Nemet, "Demand-Pull, Technology-Push, and Government-
Led Incentives for Non-Incremental Technical Change," 5 RESEARCH POLICY
700, 702 (2009); Margaret R. Taylor, Edward S. Rubin and David A. Hounshell,
"Regulation as the Mother of Innovation: The Case of SO_2 Control," 27 LAW &

the form of grants to support specific basic or applied research projects or research centers, as well as development. According to National Science Foundation data, just over $65 billion were obligated for direct non-defense related R&D funding to businesses, federal laboratories and academic institutions in 2008.[10] By comparison, 2008 industry records reported spending over $222 billion on non-defense related research and development. This figure does not include funding received from the federal government and other external sources.[11] Indirect funding includes various programs intended to reduce the cost of the emerging technology for the innovator. One well-known example of such a policy is the federal research and experimentation tax credit, initially introduced as part of the 1981 Economic Recovery Tax Act.[12] As one would expect from a tax provision, the particulars of the tax credit are complicated. Simply put, the basic credit provides for a reimbursement against a company's taxes equal to 20 percent of qualified expenditures for research and experimentation in excess of a base amount.[13] Although the design and effectiveness of the credit has been the subject of significant political and academic debate,[14] it represents significant investment by the federal government in support of the innovation process. In fiscal year 2008, it amounted to $7.1 billion, ranking it as the fourth largest corporate tax expenditure;[15] that is, an expenditure made by the government in support of an economic or other policy through reduction of an entity's tax burden.[16]

POLICY 348, 349 (2005). Other technology-push programs are designed instead to clear the way for development and diffusion of emerging technologies by removing or mitigating legal, market, or informational obstacles. Id.

[10] National Science Board, SCIENCE AND ENGINEERING INDICATORS 2012 4-35, 4-36 (2012).

[11] Id. at 4-21, 4-22.

[12] Pub.L. 97-34, 95 Stat. 172 (1981).

[13] United States Government Accountability Office, THE RESEARCH TAX CREDIT'S DESIGN AND ADMINISTRATION CAN BE IMPROVED 4–7 (2009).

[14] See Robert D. Atkinson, "Expanding the R&E Tax Credit to Drive Innovation, Competitiveness and Prosperity," 32 J. TECH. TRANSFER 617 (2007) (discussing debates over the effectiveness of the tax credit).

[15] Laura Tyson and Greg Linden, THE CORPORATE R&D TAX CREDIT AND U.S. INNOVATION AND COMPETIVENESS 35–6 (2012).

[16] Stanley S. Surrey, "Tax Incentives as a Device for Implementing Government Policy: A Comparison with Direct Government Expenditures," 83 HARV. L. REV. 705, 706 (1970).

Second, the government pursues technology advancement through "demand-pull" strategies aimed at increasing market demand for an emerging technology.[17] One demand-pull approach involves direct or indirect payments to manufacturers, vendors or end-users of new technologies that are designed to offset the increased costs of adoption. These payments can take many forms, such as low interest loans for manufacturers, tax benefits for manufacturers and end users, or subsidies for end users.[18] The government also creates demand for technology through regulatory programs that are focused upon non-economic goals, such as environmental protection and occupational health and safety.[19] Such programs typically depend upon technological innovation to achieve increasingly more demanding health standards and to deal with critical issues such as climate change. Regulations can influence technological change by creating a market for new or improved technologies in pollution control, cleaner production, remediation and other areas of concern to the policy makers.

Different regulatory designs may influence different segments of the innovation–development process. Many environmental regulations set standards by reference to existing, commercially available technology. For example, emission standards for hazardous air pollutants established under the Clean Air Act reflect levels achievable by the best performing control technologies in the relevant industry sector.[20] Regulated entities will then enter the market to purchase technologies capable of achieving those standards. Reliance on existing reference technologies thus tends to primarily drive diffusion of commercialized state-of-the-art technology. Alternatively, "technology forcing" regulation (that is, regulation setting standards that will require new technologies for compliance) include impacts further upstream in the innovation–development process. As a practical matter, however, policy makers are less likely to engage in

---

[17]   Nemet, supra note 9, at 701–2.

[18]   Timothy F. Malloy and Peter J. Sinsheimer, "Innovation, Regulation and the Selection Environment," 57 RUTGERS L. REV. 183, 224–8 (2004).

[19]   Taylor et al., supra note 9, at 372; Malloy, Regulating by Incentives, supra note 1, at 549–50.

[20]   Timothy F. Malloy, "The Social Construction of Regulation: Lessons from the War on Command and Control," 58 BUFF. L. REV. 267, 313–20 (2010).

technology forcing, and when they do, will often act when potential technologies are nearing or in the development stage.[21]

As the above discussion illustrates, the federal government is heavily involved in promoting technological advancement. In some instances, its technology policy is agnostic in terms of the subject of the research and development. The research and experimentation tax credit is one example of this neutral approach. Other efforts target innovation in specific areas, such as a loan guarantee program for renewable energy technologies. In both of those examples, the express purpose of the intervention is to drive the innovation–development process. The government as regulator also drives that process, although technological change in that context is not itself always the motivating force behind the regulatory programs. Yet even in that case, the government is seeking technological change of a particular type, such as advances in emission control techniques or adoption of safer production processes. Such targeted efforts, in particular, raise the issue of unintended adverse consequences.

6.3   UNINTENDED CONSEQUENCES: THE CASE OF MTBE

Examples of unintended consequences abound in policy making; attempts at solving one problem often generate unexpected side effects.[22] In the environmental arena, one prominent exemplar is the prohibition of one chemical or practice leading to an alternative that raises equally or

---

[21]   Id. at 550; Nicholas A. Ashford and George R. Heaton, Jr., "Regulation and Technological Innovation in the Chemical Industry," 46 L. & CONTEMP. PROBS. 109, 127 (1983).

[22]   Bryan D. Jones & Frank R. Baumgartner, THE POLITICS OF ATTEN-TION 14 (2005) ("Solving public problems is neither simple nor straightforward. Each solution may have multiple implications for the problem and various "side consequences" going beyond the first problem and potentially creating new issues to deal with. ... And these side effects can be uncertain as well."). Unintended consequences appear in a broad range of policy contexts. See Cass R. Sunstein, "Political Equality and Unintended Consequences," 94 COLUM. L. REV. 1390 (1994) (discussing campaign finance law); Ryan Miske, "Can't Cap Corporate Greed: Unintended Consequences of Trying To Control Executive Compensation Through the Tax Code," 88 MINN. L. REV. 1673 (2004) (discussing tax law); A.A. Sommer, Jr., "Preempting Unintended Consequences," 60 LAW AND CONTEMP. PROB. 231 (1997) (discussing securities law).

more troubling outcomes.[23] Despite the ubiquitous references to unintended consequences in legal scholarship and in public policy discourse, few users of the term have sought to define its contours.[24] Sociologist Robert Merton is a notable exception. In his seminal paper, "The unanticipated consequences of purposeful social action," Merton parsed the concept into its component parts, and identified a set of sources of unintended consequences that has significant continued relevance today.[25] I follow Merton's example and limit unintended consequences to those that are "unanticipated."[26] A consequence is unanticipated if it is unforeseen by the relevant decision maker, without regard to whether the actor should have or could have foreseen it.[27] For these purposes, an outcome is foreseen if the decision maker either expected the outcome or recognized that it was likely. Outcomes that are considered, but incorrectly rejected as being low probability risks, would be unanticipated.

A complete or even ambitious partial survey of technology policies that have resulted in unintended consequences is beyond the scope of this chapter. Nonetheless, one example can provide a sense of the problem. The story of MTBE is a useful example of unintended consequences flowing from a demand-pull scenario. MTBE is a volatile organic compound produced as a by-product of petroleum refining.[28] First produced commercially in the United States in 1979, MTBE was added to mid- to high-grade gasoline at low levels to reduce "knocking."[29] But

---

[23]   Molly M. Jacobs, Joel Tickner& David Kriebel, LESSONS LEARNED SOLUTIONS FOR WORKPLACE SAFETY AND HEALTH 85 (2011).

[24]   See Richard Vernon, "Unintended Consequences," 7 POLITICAL THEORY 57, 71 (1979) (concluding that "different theorists have had different things in mind when dealing with unintended consequences, some having had at least two in mind at once without distinguishing between them"); Sommer, Jr., supra note 21, at 231 ("Who enacted this law, who enforces it, and its exact scope are obscure").

[25]   Robert K. Merton, "The Unanticipated Consequences of Purposeful Social Action," 1 AM. SOC. REV. 894, 897 (1936).

[26]   Id. at 894. Alternatively, one could define intent by reference to expectations. In other words, an outcome is intended if it is reasonably anticipated to occur as a result of the course chosen.

[27]   A consequence can be beneficial, detrimental or neutral in the view of the policy maker. Of course, the same outcome could be both detrimental and beneficial to one interested party, or have different impacts on distinct parties.

[28]   HANDBOOK OF MTBE AND OTHER GASOLINE OXYGENATES 2 (Halim Hamid and Mohammad Ashraf Ali, eds, 2004).

[29]   65 Fed. Reg. 16094, 16098 (March 24, 2000); Thomas O. McGarity, "MTBE: A Cautionary Tale," 28 HARV. ENVT. L. REV. 281, 284 (2004).

its status as an "oxygenate" – a hydrocarbon containing oxygen – would catapult it to fame and ultimately to infamy.

Two separate provisions of the 1990 amendments to the Clean Air Act created robust demand for MTBE in the 1990s. First, Section 187(b) generally mandated wintertime use of oxygenated fuels (or oxyfuel) to reduce carbon monoxide (CO) emissions in areas of serious nonattainment with the National Ambient Air Quality Standard for CO.[30] Second, Section 211(k)(2)(B) pressed oxygenates into service in the battle against toxic automotive emissions that were associated with aromatic compounds such as benzene, toluene and xylene found in gasoline.[31] That section required reformulation of gasoline by which the aromatics would be displaced by ostensibly benign compounds like ethanol or MTBE.[32] Although neither the oxyfuel nor reformulated gasoline provisions required the use of MTBE specifically,[33] the legislative history demonstrates that the leading players in the Senate and House fully expected MTBE (along with ethanol) to be the oxygenate of choice for industry and regulators alike, particularly for the oxyfuel program.[34] Indeed, in the

---

[30] National Science And Technology Council, INTERAGENCY ASSESSMENT OF OXYGENATED FUELS 1-5–1-8 (June 1997).

[31] 136 CONG. REC. S3504 (March 29, 1990) (statement of Senator Harkin).

[32] Clean Air Act Amendments of 1990, Pub. L. No. 101-549, § 211(k)(2)(B) (1990).

[33] The legislative history of both provisions supports this notion of "fuel neutrality." See Exxon Mobil Corp. v. US Envtl. Prot. Agency, 217 F.3d 1246, 1253 (9th Cir. 2000) (stating "The legislative history suggests that fuel neutrality on the part of the Administrator was a goal of the provisions").

[34] See, for example, 136 Cong. Rec. S3504, S3511 (March 29, 1990) (statement of Sen. Daschle) ("One of the most important attributes of a clean octane program is that it can be implemented immediately. In fact, there are 11 EPA-approved octane-enhancers that can be used to replace aromatics, including MTBE, ETBE, ethanol and other oxygenates. The ethers, especially MTBE and ETBE, are expected to be major components of meeting a clean octane program."); 1990 Conf. Rept. (October 26, 1990) ("The agreement establishes an oxygen content level of 2.7 percent in 44 cities with carbon monoxide pollution, starting in 1992. These provisions will encourage the use of oxygen-containing additives like ethanol and MTBE, a natural gas derivative"); 136 Cong. Rec. S16954 (1990) (remarks of Sen. Chafee noting that the reformulated fuels program "will encourage the use of oxygen-containing additives like ethanol and MTBE"); 136 Cong. Rec. S17514 (1990) (remarks of Sen. Heinz observing that "[t]he requirements for reformulated gasoline will also encourage the use of oxygen-containing additives like ethanol and MTBE").

Senate, the original version of the oxyfuel provisions was amended in the conference committee to ease the way for use of MTBE.[35]

The use of MTBE as an oxygenate under the freshly minted 1990 Clean Air Act Amendments opened the door to significant market growth. While the oxyfuel and reformulated gasoline provisions were nominally neutral, MTBE had significant advantages over other oxygenates, most notably ethanol, in that it was relatively inexpensive, and could be easily blended with gasoline at the refinery and transported through existing pipelines.[36] In the afterglow of the passage of the Clean Air Act Amendments, production of MTBE in the United States doubled.[37] By the year 2000, MTBE was one of the leading end-use high volume chemicals produced in the United States.[38] Yet shortly after the 1992 kickoff of the oxyfuel program, trouble for MTBE appeared on the regulatory horizon.

Oxyfuels containing MTBE were introduced in CO nonattainment areas across the country in the winter of 1992. Almost immediately, regulators received complaints from motorists in Alaska, New Jersey and other locales regarding the perceived effects of MTBE fumes, including

---

[35] McGarity, supra note 29, at 309.

[36] Pamela M. Franklin, "Clearing the Air: Using Scientific Information to Regulate Reformulated Fuels," 34 ENVTL. SCI. TECH. 3857, 3858 (2000); David E. Stikkers, "Octane and the Environment," 299 SCI. OF THE TOTAL ENV'T 37, 50 (2002).

[37] US Department of Energy, Energy Information Administration, Petroleum Supply Monthly, Washington, DC, January 2003 and historical editions. The increase was tied to both the oxyfuel provisions of Section 187 and the reformulated gasoline program under Section 211(k). The Energy Information Administration of the Department of Energy reported that "the oxygenated gasoline program stimulated an increase in MTBE production between 1990 and 1994. MTBE demand increased from 83,000 in 1990 to 161,000 barrels per day in 1994. The reformulated gasoline (RFG) program provided a further boost to oxygenate blending. The MTBE contained in motor gasoline increased to 269,000 barrels per day by 1997." Energy Information Administration, MTBE, Oxygenates, and Motor Gasoline, available at http://www.eia.doe.gov/emeu/steo/pub/special/mtbe.html (accessed July 16, 2013).

[38] Stikkers, supra note 36, at 50; Martin Krayer von Krauss and Poul Harremoes, "MTBE in Petrol as a Substitute for Lead," in THE PRECAUTIONARY PRINCIPLE IN THE 20TH CENTURY 121, 122 (Paul Harremoes, David Gee, Malcom MacGarvin, Andy Stirling, Jane Keys, Brian Wynne and Sofia Guedes Vaz, eds, 2002) (noting that in 1995, MTBE was the third most produced organic chemical in the United States).

nausea, headaches and asthma attacks.[39] By 1994, Alaska had banned MTBE as an oxygenate in response to the alleged inhalation effects, and New Jersey likewise took action to limit its use.[40] By all accounts, however, the beginning of the end for this particular oxygenate can be traced to the 1996 discovery in MTBE in Santa Monica, California's groundwater on the heels of the introduction of reformulated gasoline. Soon thereafter, the trickle of MTBE contamination reports became a flood. By 2000, individual states began to phase out MTBE use, with the federal government following suite in 2005.[41]

6.4   UNINTENDED CONSEQUENCES: THE SOURCES

How is it that after years of hearings, consideration and debate over the problem of nonattainment and toxics, Congress set in motion not one but *two* programs that relied so heavily on MTBE?[42] In answering this question, I begin by considering a set of factors that contribute to unintended consequences more generally. The set is derived from explanations for unintended consequences offered by various commentators, which serve as a useful starting point for our analysis.[43] Because this

---

[39]   Stikkers, supra note 35 at 51; Michael Davis and William H. Farland, "The Paradoxes of MTBE," 61 TECH. SCI. 211, 212–16 (2001).

[40]   Serap Erdal & Bernard D. Goldstein, "Methyl Tert-Butyl Ether as a Gasoline Oxygenate: Lessons for Environmental Public Policy," 25 ANN. REV. ENERGY ENVTL. 765, 768 (2000).

[41]   California's governor Gray Davis acted first, issuing an Executive Order in 1999 that provided for a phase-out to be completed by 2002.

[42]   By "Congress" here and throughout the article, I refer to the subset of representatives and senators actively and substantively involved in the formulation of and deliberation over these provisions of the Clean Air Act Amendments.

[43]   See Merton, supra note 25, 898–904 (discussing the stochastic nature of human behavior, complexity, ignorance of ascertainable facts, rational ignorance, error, "imperious immediacy of interest" (focus on short-term consequences), fundamental values driving choices, and the effect of predictions and policies on the occurrence of the predicted behavior); John R. Graham and Jonathan B. Wiener, RISK VS. RISK: TRADEOFFS IN PROTECTING HEALTH AND THE ENVIRONMENT 228–41 (1995) (identifying omitted voice, heuristics, bounded oversight roles, new-old bias, and behavioral responses). Although Merton did not use the recently popularized term "heuristics," his discussions of human behavior, error and imperious immediacy of interest presage Graham and Wiener's explication of the impact of heuristics on unanticipated consequences; likewise, Merton's notion of the stochastic nature of human behavior mirror Graham and Wiener's concept of "behavioral responses." Id.

chapter focuses upon decision making by policy makers (and legislators in particular), I add to those explanations the dynamic and uncertain nature of knowledge[44] and the unavoidably political nature of policy formulation. The resulting set of factors can be organized into three categories: attention- and information-processing barriers, inherent unforeseeability, and political influence. They are described in more detail in Table 6.1.

In the case of MTBE and the Clean Air Act Amendments, the attention/information category seems most salient in explaining the unintended consequences. The second two categories did not appear to play a substantial role in generating unintended consequences. The case did not raise problems with respect to the random nature of human behavior, nor did it reflect the characteristics of a complex system defying prediction. The political influence category certainly appears relevant at first glance, in that the sophisticated, well-funded oil industry interest clearly had strong motivation to advance MTBE's place in air quality regulation. However, the formal record contains no clear evidence of political motivations driving Congress' actions, apart from its potential role in the delegation of some issues to EPA jurisdiction discussed below.

Regarding the attention/information category, it does appear from the record that decision makers in Congress, particularly members of the relevant environmental committees in the House and Senate, lacked complete knowledge of *existing* information regarding many of the potential adverse impacts of ramping up MTBE usage. One glaring example is the risk of pervasive groundwater contamination from service stations that stored gasoline with MTBE. This information was not communicated to Congress, despite the fact that there was a significant amount of such information known to both industry and EPA. Prior to 1990, the refining industry had collected information regarding the health and environmental effects of MTBE as part of a coordinated research

---

44   Kenneth Arrow recognized the role of uncertainty in the political environment: "It is my view that most individuals underestimate the uncertainty of the world. This is almost as true of economists and other specialists as it is of the lay public. ... When developing policy with wide effects for an individual or a society, caution is needed because we cannot predict the consequences." Kenneth Arrow, "I Know a Hawk from a Handsaw," in EMINENT ECONOMISTS: THEIR LIFE PHILOSOPHIES 46 (Michael Szenberg, ed., 1992).

Table 6.1   Sources of Unintended Consequences

Category	Factor	Description
Attention/ information barriers	Rational ignorance	The expected costs of obtaining additional information exceed the expected value to the decision maker of having the information.
	Uncertain/ incomplete info	The information available to the decision maker and participants in the decision making process is uncertain or incomplete, and cannot be further developed.
	Asymmetric information	Information relevant to the decision exists, but is withheld from the decision maker by another participant in the decision making process.
	Omitted voices	Parties with information relevant to the decision are excluded from the decision making process intentionally or inadvertently.
	Fragmentation	Within an organization or institution, information relevant to a decision fails to flow from one sub-unit to another due to deficiencies in organizational/institutional structure or operation.
	Heuristics/cognitive biases and strategies	Unconscious biases in information processing or decision making impair the decision-maker's recognition of or attention to potential consequences of a decision. For example, the availability heuristic may lead the decision maker to consider only those potential consequences with which they are familiar.[45] Motivated reasoning may lead a decision maker to undervalue or even discount information that is contrary to the decision maker's desired outcome.[46]

---

[45] Paul Slovic, Bruce Fischhoff and Sarah Lichtenstein, "Perceived Risk: Psychological Factors and Social Implications," 376 PROC. R. SOC. A 17, 29–31 (1981).

[46] Ziva Kunda, "The Case for Motivated Reasoning," 108 PSYCHOL. BULL. 480 (1990).

Table 6.1 Continued

Category	Factor	Description
Inherent unforsee-ability	Unpredictable human behavior	Human behavior is often unpredictable; individuals and groups sometimes act in ways that are inconsistent with prior behavior.
	Inherent system complexity	Policy decisions attempt to predict the reactions of complex natural, technical and human systems. Complex systems by nature are often unpredictable in that ostensibly small actions can have unexpectedly significant impacts on system behavior in ways that are extremely difficult to model or predict in advance.[47]
Political influence	Ideological constraints	Basic ideology or values may lead a decision maker to exclude certain issues or consequences from consideration in generating or evaluating potential policies.
	Political motivation	The decision maker avoids consideration of potential adverse consequences for political reasons. For example, in the interest of achieving a negotiated outcome, the decision maker defers consideration of a potential issue to a later time. Also to avoid responsibility for adverse outcomes, the decision maker shifts responsibility for the risky portion of the decision to another party.

program supervised by the American Petroleum Institute (API).[48] The group was aware that MTBE groundwater contamination was already occurring, that MTBE migrated more quickly than other gasoline constituents, and that water with low concentrations of MTBE had a pungent odor and turpentine-like taste. The group had conducted some toxicology testing regarding health effects from inhalation, but resisted internal recommendations for testing for effects from ingestion.[49]

Various offices in EPA were also aware of MTBE concerns regarding inhalation and groundwater exposure. EPA had received reports of MTBE contamination in groundwater in North Carolina by 1986,[50] and

---

[47]   National Research Council, DECISION MAKING FOR THE ENVIRONMENT: SOCIAL AND BEHAVIORAL SCIENCE RESEARCH PRIORITIES 19 (2005).
[48]   McGarity, supra note 29, at 298–300.
[49]   Id. at 298–99.
[50]   51 Fed. Reg. 41417 (November 14, 1986).

in New Jersey, Maine and New Hampshire by 1988.[51] As the reports regarding MTBE contamination diffused through the agency, a variety of offices within EPA acknowledged the problem and responded formally. First up was the Office of Toxic Substances, which initiated efforts in late 1986 to mandate health effects testing of MTBE under the Toxic Substances Control Act.[52] While those efforts were driven in large part by concerns over inhalation of MTBE by workers and consumers, EPA noted that "[a]lthough only a few cases of ground water contamination are currently documented, the rapid growth in production, transport, and use of MTBE will probably contribute to an increase in incidents of contamination."[53] Those efforts culminated in a consent order signed by five MTBE manufacturers on January 22, 1988, which, in response to industry arguments, specifically excluded mention of groundwater ingestion.[54] On that very same day, EPA's Office of Drinking Water added MTBE to its drinking water priority list, based upon its "potential for widespread contamination."[55] The Office of Underground Storage Tanks joined the chorus later that year, noting in a guidance document that "MTBE is easily transported by groundwater away from a spill site."[56]

Despite the availability of this information to industry and some parts of EPA, research reveals only one reference to the potentially adverse health and environmental effects of MTBE in the numerous hearings, committee reports, and floor debates concerning the Clean Air Act Amendments between 1987 and 1990. That one exception occurred in November of 1987, when the Senate Committee on Environment and Public Works reported on the Clean Air Standards Attainment Act of 1987, which included a provision mandating the wintertime use of oxygenated fuels in CO nonattainment areas.[57] The Committee's report acknowledged that fuel additives "can present a significant threat to

---

[51]   53 Fed. Reg. 10391 (March 31, 1988).

[52]   Id.

[53]   Id.

[54]   Id.

[55]   53 Fed. Reg. 1892 (January 22, 1988). Under the Safe Drinking Water Act, EPA was required to create a list of contaminants which "may have any adverse effect on the health of persons and are known or anticipated to occur in public water systems and which may require regulation under the Act." Safe Drinking Water Act, § 1412(b)(3).

[56]   EPA, Cleanup of Releases from Petroleum USTs: Selected Technologies 10 (EPA/530/UST-88/001 April 1988).

[57]   S. REP. NO. 100–231 at 3502 (1987). A subsequent report from that same committee reporting out a later version of the bill included similar language regarding MTBE. S. REP. NO. 101–228 at 8457 (1989).

public health and the environment," and observed that EPA had recently "proposed a voluntary testing program (to be conducted by the manufacturers) on the health effects of methyl tri-butyl ether or MTBE."[58] The voluntary testing program mentioned in the report is apparently a reference to the consent order issued by EPA under TSCA, which required very little toxicological testing with respect the groundwater ingestion route.[59] Nonetheless, the full committee rejected a proviso in the subcommittee bill mandating "thorough and complete health and environmental effects testing" within five years of all fuel additives used in gasoline.[60] Instead the Committee expressed its strong expectation that EPA would use existing authorities under the Clean Air Act to "conduct a thorough review of fuel additives in the near-term."[61]

It appears, therefore, that Congress had sparse information regarding the risks of MTBE and took little action on the limited information it did receive. Of the six mechanisms by which attention/information barriers operate, three appear to be most relevant here: (1) asymmetric information, (2) fragmentation, and (3) cognitive biases and strategies. The most obvious mechanism is asymmetric information. Both industry members and EPA had greater access to information regarding the potential impacts, but they apparently did not share much of that information with decision makers within the legislature. The oil industry's reluctance to share negative information about MTBE is hardly surprising; its members had a great deal to gain from expanded markets for the chemical.

As for EPA, there are several possible explanations for the trickle of information reaching Congress with respect to MTBE. Some commentators argue that fragmentation within the agency itself may have been a significant contributing factor. The offices most engaged in collecting data with respect to MTBE's risks – the Office of Toxic Substances, Office of Drinking Water, and Office of Underground Storage Tanks – were not significantly involved in the Clean Air Act reform efforts in Congress. There did not appear to be formal mechanisms ensuring that such information flowed to the Office of Mobile Sources (OMS), which

---

[58] Id.
[59] Testing Consent Order on Methyl Tert-Butyl Ether and Response to Interagency Testing Committee, 53 Fed. Reg. 10391 (March 31, 1998); McGarity, supra note 28, at 301.
[60] Sen. Rept. 100-231 at 3503 (November 20, 1987).
[61] Id.

was involved in the legislative activity.[62] While this is plausible, there is evidence suggesting that the OMS interacted with the other offices during this period, and thus may have been aware of the groundwater issues. In a September 1989 report provided to Congress regarding methanol as an alternative fuel, OMS engaged in extensive discussion of groundwater issues associated with methanol spilled or leaking from underground storage tanks.[63] Given the specificity of that water quality impact and underground storage tank discussion, it is likely that the OMS engaged in some communication with the Office of Drinking Water. Whether those lines of communication also functioned with respect to MTBE is unclear, but the methanol experience at least demonstrates that fragmentation was less pervasive than some commentators suggest.

Another possible explanation for EPA's apparent silence on the groundwater issue may be selective perception, a cognitive phenomenon in which individuals' training and experience bias the manner in which they perceive and act on information.[64] While selective perception can affect various aspects of the decision making process, its most relevant effect for our purposes in on problem identification. Here, whether an individual frames a situation as problematic – and thus worthy of attention – is dependent upon the individual's knowledge, goals and values.[65] Was potential groundwater contamination from increased MTBE use as an oxygenate a "problem"? Perhaps not, if one views MTBE as just another component of gasoline, and one of the less toxic ones to boot. No doubt releases from underground storage tanks were of concern, but new underground storage tank regulations were being implemented to address the broader issue. Such a framing of the impacts of MTBE could lead agency staff and managers to dismiss groundwater contamination as a "problem" for purposes of the Clean Air Act. Research reveals no contemporaneous evidence regarding EPA's failure to press the groundwater contamination issue before Congress, although the Congressional

---

[62] Franklin., supra note 36, at 3860–61; Erdal and Goldstein, supra note 40, at 796–7; but see McGarity, supra note 29, at 317.

[63] Office Of Mobile Sources, Environmental Protection Agency, ANALYSIS OF THE ECONOMIC AND ENVIRONMENTAL EFFECTS OF METHANOL AS AN AUTOMOTIVE FUEL 69–70 (1989).

[64] Ronald M. Bradfield, "Cognitive Barriers in the Scenario Development Process," 10 ADVANCES IN DEVELOPING HUMAN RESOURCES 198, 204–5 (2008).

[65] Id. at 205; Linda M.H. Lai and Kjell Grønhaug, "Managerial Problem Finding: Conceptual Issues and Research Findings," 10 SCANDINAVIAN J. MGMT. 1, 4–5 (1994).

Research Service's assessment of EPA's response to the problems that emerged later suggests that selective perception could have been at work:

> In EPA's view, studies to date have not indicated that MTBE poses any greater risk to health than other gasoline components, such as benzene. As a result, the Agency has chosen to respond by providing information, intensifying research, and monitoring implementation of [Underground Storage Tank] requirements discussed earlier.[66]

Even if we can explain industry's and EPA's limited disclosures to Congress, what are we to make of Congress' inattention to the risks that *were* disclosed? Clearly, some in Congress knew of the MTBE testing under TSCA. One possible response rejects the notion that Congress failed to act in anticipation of potential problems; after all, Congress did call for continued evaluation of fuel additives like MTBE by EPA. The legislative history of the oxyfuel provision reflects Congress' understanding that EPA would scrutinize oxygenate additives for unacceptable health and environmental impacts under Section 211 of the Clean Air Act, and would take appropriate action.[67] Likewise, regarding reformulated gasoline, the amendments specifically directed EPA to consider environmental impacts broadly defined in establishing standards for the new type of fuel.[68]

That argument sputters badly when the context of the legislative and regulatory processes are considered. With regard to the oxyfuel program, EPA had limited authority under the Clean Air Act to restrict the use of MTBE as a fuel additive. While EPA could require health testing of fuel additives like MTBE under Section 211(b) of the Act, its ultimate power to restrict their use was limited to situations in which the *emission*

---

[66]   James E. Mccarthy and Mary Tiemann, MTBE IN GASOLINE: CLEAN AIR AND DRINKING WATER ISSUES (March 24, 1988).

[67]   See S. REP. NO. 101-228 (Dec. 20, 1989) ("The Agency should implement the authorities contained in section 211 with respect to fuel additives by developing criteria (based on factors such as emissions of the additive or combustion byproducts, impact of the additive or byproducts on the emissions control systems of motor vehicles, exposure to the additive, byproducts or atmospheric transformation products and the potency of the additive, byproduct or transformation products as human and environmental toxicants) for testing determinations. Each additive should then be evaluated according to the criteria and testing programs for additives of concern should be commenced using the authority of section 211(b).")

[68]   Clean Air Act § 211(k).

products of the fuel additive endangered public health.[69] Thus, the dangers to public health from groundwater contamination or resulting from inhalation at the pump were not even a relevant factor. Moreover, based on the well-documented, very public institutional failures on EPA's part to implement testing and intervention authorities in the past, Congress had little reason to believe that the agency would evaluate currently used additives in a timely and complete manner. One striking example is the fuel additive testing provision itself. Although that provision was enacted as part of the 1970 Clean Air Act, some 20 years later EPA had yet to establish the testing program.[70] As the legislative

---

[69] §§ 211(a)–(b) established a registration and testing program for fuels and fuel additives. Under § 211(c), the Administrator of EPA could "control or prohibit the manufacture, introduction into commerce, offering for sale, or sale of any fuel or fuel additive for use in a motor vehicle, motor vehicle engine, or nonroad engine or nonroad vehicle (A) if in the judgment of the Administrator any emission product of such fuel or fuel additive causes, or contributes, to air pollution which may reasonably be anticipated to endanger the public health or welfare, or (B) if emission products of such fuel or fuel additive will impair to a significant degree the performance of any emission control device or system which is in general use, or which the Administrator finds has been developed to a point where in a reasonable time it would be in general use were such regulation to be promulgated." Id. at § 211(c).

[70] EPA explained the status of its testing program development efforts in an advance notice of proposed rulemaking in 1992: "The 1970 CAA also provided EPA with discretionary authority to establish additional requirements for fuel and fuel additive registration. According to section 211(b)(2), EPA 'may also require' producers 'to conduct tests to determine potential public health effects of such fuel[s] or additive[s] (including, but not limited to, carcinogenic, teratogenic, or mutagenic effects).' ... EPA did not exercise its discretionary authority to require fuel and additive testing under section 211(b)(2) when the general registration regulations under section 211(b)(1) were issued in 1975. However, in the CAA Amendments of 1977 (Pub. L. 95-5, August 7, 1977), Congress added section 211(e) to the statute, which made implementation of section 211(b)(2) mandatory and contained additional provisions concerning the implementation of the statute. Section 211(e)(1) requires implementation of the section 211(b)(2) authority within one year of enactment of the 1977 amendments. In an effort to fulfill this requirement, EPA published an Advanced Notice of Proposed Rulemaking (ANPRM) in 1978 (see 43 Fed. Reg. 38607, August 29, 1978; Docket ORD-78-01); however, the rulemaking did not go forward during the next ten years. Nevertheless, this action has remained on EPA regulatory agenda, and a development plan for the rulemaking was created in 1988," 57 Fed. Reg. 13168 (April 15, 1992).

history of the Clean Air Act Amendments demonstrates, Congress was painfully aware of EPA's shortcomings in this and numerous other areas.[71]

In contrast to the fuel additive testing provision, the reformulated gasoline provisions clearly contemplated that the agency would consider a broad range of potential consequences associated with specific formulations of the gasoline, including oxygenates. In fact, Section 211(k)(1) requires that EPA must consider "the cost of achieving such emission reductions, *any nonair-quality and other air-quality related health and environmental impacts* and energy requirements.[72] This type of broad-minded mandate appears in numerous other Clean Air Act programs.[73] While no systematic study of EPA's implementation of this mandate over time and across programs appears to be available, a review of randomly selected Federal Register preambles indicates that the agency typically engaged in a relatively superficial evaluation of cross-media impacts (that is, consequential effects on water quality, solid waste management, etc.).[74] The 211(k)(1) analysis ultimately completed by EPA in December 1993 reflected this same implementation style; it focused primarily on

---

[71]   See, for example, S. REP. 101–228 at 3503 (having specifically noted the need to evaluate the safety of fuel additives such as MTBE, the report concludes: "As the Agency has not implemented already enacted non-discretionary provisions of law in this respect, this activity is to be completed on an expedited schedule."); 136 Cong. Rec. S3504, S3512 (March 29, 1990) (statement of Sen. Daschle) (discussing reduction of aromatics in gasoline, he noted that "[u]nfortunately, EPA has known about this problem for more than a decade and has repeatedly failed to address it. Despite the fact that mobile source toxics account for more than half the air pollution-related cancer deaths annually, EPA does not regulate mobile source toxics. I simply do not trust EPA to implement a program that has been stonewalled for more than a decade.").

[72]   Clean Air Act § 211(k)(1) (emphasis added).

[73]   Clean Air Act § 111(a)(1) (requiring consideration of any non-air quality health and environmental impact and energy requirements in setting standards of performance for new sources); Clean Air Act § 112(d) (requiring consideration of "any nonair quality and other air quality related health and environmental impacts, and energy requirements" in promulgating standards for sources of hazardous air pollutants); Clean Air Act § 169A(g)(1) (mandating consideration of "the costs of compliance, the time necessary for compliance, and the energy and nonair quality environmental impacts of compliance, and the remaining useful life of any existing source" in establishing regulations regarding the control of regional haze in certain national parks and wilderness areas.")

[74]   See, for example, 45 Fed. Reg. 78174 (November 25, 1980) (briefly reviewing water quality and solid waste impacts of an air pollution control standard); 63 Fed. Reg. 50280, 50287 (September 21, 1998) (setting out cursory

cost and other air-quality related health impacts, making no mention of MTBE as a either a groundwater contaminant or an inhalation concern.[75]

On balance, Congress seemed to engage in classic "pass the buck" behavior with respect to the MTBE risks of which at least some members seemed to be aware.

This behavior may be traceable to a cognitive coping strategy used to avoid the psychological stress associated with making a difficult decision.[76] Such stress flows from a variety of factors, the most salient here being concern over accountability for a decision which turns out to have negative outcomes. The stress can be reduced by shifting the decision to an agent – in this case the regulatory agency.[77] Alternatively, in the legislative setting, the decision to delegate may simply be the result of political motivation, one aspect of the political influence category. The delegation could be a rational reaction to the political impacts that undesirable impacts of a decision may have in the future.[78] In the case of MTBE, Congress could take credit through delegation for the decision to use oxygenates to reduce harmful emissions, while placing the responsibility (and consequently, the blame) for the choice of the specific oxygenate on the regulatory agency. Whatever the root of Congress' avoidance of the groundwater issue, it clearly contributed to the unintended consequences of the Clean Air Act. The question remains whether formal technology assessment can prevent such situations in Congress from going forward.

---

review of air, water and solid waste impacts of standard applicable to emissions of hazardous air pollutants from pharmaceuticals production).

[75]   See Office Of Mobile Sources, Environmental Protection Agency, FINAL REGULATORY IMPACT ANALYSIS FOR REFORMULATED GASOLINE 301 (EPA420-R-93-017, 1993) ("Since it has been determined that the health and environmental impacts of such fuel reformulations are beneficial and increase as the percentage emission reduction increases, and that the energy impacts are minimal, the Phase II reformulated gasoline performance standards are determined based primarily on cost-effectiveness.")

[76]   Jane Beattie, Jonathan Baron, John C. Hershey and Mark D. Spranca, "Psychological Determinants of Decision Attitude," 7 J. BEHAV. DECISION MAKING 129, 130–1 (1994); Werner F. M. De Bondt and Richard H. Thaler, FINANCIAL DECISION-MAKING IN MARKETS AND FIRMS: A BEHAVIORAL PERSPECTIVE 8–9 (Nat'l Bureau Econ. Res., Working Paper No. 4777, 1994).

[77]   Beattie, supra note 76, at 133.

[78]   Stefan Voigt and Eli M. Salzberger, "Choosing Not To Choose: When Politicians Choose To Delegate Powers," 55 KYKLOS 289, 294–5 (2004).

6.5 TECHNOLOGY ASSESSMENT

This section begins with a brief overview of the nature and types of technology assessment, followed by examination of the benefits and limits of technology assessment in the legislative setting. Technology assessment has been defined as "an applied process that considers the societal implications of technological change in order to influence policy to improve technology governance."[79] This admittedly broad definition captures the essence of the process, while leaving room for its many forms and numerous methodologies. For example, while our focus here is the legislative setting, technology assessment is also used by regulatory agencies, non-governmental institutions, and industry alike.[80] Generally speaking, there are three types of technology assessment:

1.   Traditional technology assessment: One of the first forms of formal technology assessment, this approach typically serves as an early warning function regarding the potential impacts of technologies for policy makers.[81] It relies primarily, if not exclusively, upon analysis performed by technical experts.[82] Initially focused upon economic and technological impacts, over time it expanded in some applications to include environmental, social and cultural impacts. Participation of stakeholders and the general public is passive,

---

[79]   A. Wendy Russell, Frank M. Vanclay and Heather J. Aslin, "Technology Assessment in Social Context: The Case for a New Framework for Assessing and Shaping Technological Development," in IMPACT ASSESSMENT AND PROJECT APPRAISAL 109 (2010); see also Laura Cruz-Castro and Luis Sanz-Menendez, "Politics and Institutions: European Parliamentary Technology Assessment," 72 TECH. FORECAST. SOC. CHANGE 429, 430 (2005) ("the production of information about the possible consequences of S&T developments to improve public policies"); Joseph F. Coates, "A 21st Century Agenda for Technology Assessment," 67 TECH. FORECAST. SOC. CHANGE 139, 139 (2001) (defining it as "a policy study designed to better understand the consequences across society of the extension of the existing technology or the introduction of a new technology with emphasis on the effects that would normally be unplanned and unanticipated").

[80]   Thien A. Tran and Tugrul Daim, "A Taxonomic Review of Methods and Tools Applied in Technology Assessment," 75 TECH. FORECAST. SOC. CHANGE 1396–1402 (2008).

[81]   Jan Van Den, Karel Mulder, Marjolijn Knot, Ellen Moors and Philip Vergragt, "Traditional and Modern Technology Assessment: Toward a Toolkit," 58 TECH. FORECAST. SOC. CHANGE 5, 5–6 (1998).

[82]   Josée C.M. Van Eijndhoven, "Technology Assessment: Product or Process," 54 TECH. FORECAST. SOC. CHANGE 269, 275–6 (1997).

limited to providing structured input through advisory panels and through review and comment procedures.[83]

2. Participatory technology assessment: This type emphasizes the social nature of technology by expanding upon traditional technology assessment to actively integrate outside viewpoints and public values.[84] It brings citizens into the process, providing opportunities for them to learn about the technology and to express their views through techniques such as consensus conferences, citizens' juries and planning cells.[85]

3. Constructive technology assessment: This type focuses on the earliest stages of technological change, involving scientists, regulators, workers, users and the broader public in the development and design of technology.[86] Constructive technology assessment, and variants such as "interactive" and "real-time technology assessment," uses a variety of techniques to integrate social concerns and preferences into technology development.[87]

Technology assessment emerged as an element of legislative policy making in the United States in the early 1970s with the creation of the Office of Technology Assessment (OTA). As a substantive matter, the OTA was to ensure that "the consequences of technological applications be anticipated, understood, and considered in determination of public policy on existing and emerging national problems."[88] As a political matter, it was also intended to offset the advantage that the executive branch had over Congress in terms of access to scientific and technical resources.[89] The nature and scope of technology assessments done by the OTA evolved over time, and in any given case depended upon the needs

---

[83] Albert C. Lin, "Technology Assessment 2.0: Revamping our Approach to Emerging Technologies," 76 BROOK. L. REV. 1309, 1349–1350 (2010–2011); Richard Sclove, REINVENTING TECHNOLOGY ASSESSMENT: A 21ST CENTURY MODEL 10–11 (2010).

[84] Frank M. Vanclaya, A. Wendy Russell and Julie Kimber, "Enhancing Innovation in Agriculture at the Policy Level: The Potential Contribution of Technology Assessment," 31 LAND USE POL'Y 406, 408 (2013).

[85] Lin, supra note 83, at 1350–1; Jan van den Ende et al., supra note 81, at 12.

[86] Lin, supra note 83, at 1353–4; Vanclay et al., supra note 84, at 408.

[87] Lin, supra note 83, at 1353–4; Van Eijndhoven, supra note 82, at 280.

[88] The Technology Assessment Act of 1972, Pub. L. 92-484.

[89] Gregory C. Kunkle, "New Challenge or the Past Revisited? The Office of Technology Assessment in Historical Context," 17 TECH. IN SOC'Y 175, 177–8 (1995).

identified by the congressional sponsor and the recommendations of the advisory panel created to guide the assessment.[90] While some critics have questioned the actual impact of the OTA's technology assessments on specific legislation, most commentators conclude that they played an important background role in framing issues and building capacity and knowledge.[91] A victim of political wrangling after the shift of control of Congress in 1994 to the Republicans, the OTA was de-funded and essentially abolished as of 1996.[92] Ironically, while legislative technology assessment diffused to a variety of European countries,[93] it has been largely missing from the scene in the United States since the OTA's demise.[94]

The OTA engaged in traditional technology assessment mixed, in later years, with some policy analysis. While the OTA had a well-developed process for assessment (including project planning, extensive use of advisory panels and peer review, and public participation), it never embraced existing formal methods for technology assessment.[95] Instead, it relied upon an eclectic, ad hoc approach in which the research methods and assessment methodologies were selected on a project by project

---

[90]   Roger C. Herdman and James E. Jensen, "The OTA Story: The Agency Perspective," 54 TECH. FORECAST. SOC. CHANGE 131, 135–7 (1997).

[91]   Karen Bogenschneider & Thomas J. Corbett, EVIDENCE-BASED POLICYMAKING: INSIGHTS FROM POLICY-MINDED RESEARCHERS AND RESEARCH-MINDED POLICYMAKERS 272–3 (2010); Vary T. Coates, "Technology Forecasting and Assessment in the United States: Statistics and Prospects," in 21ST CENTURY OPPORTUNITIES AND CHALLENGES: AN AGE OF DESTRUCTION OR AN AGE OF TRANSFORMATION 8, 18 (Howard F. Didsbury, Jr., ed., 2003); Bruce Bimber, THE POLITICS OF EXPERTISE IN CONGRESS: THE RISE AND FALL OF THE OFFICE OF TECHNOLOGY (1996).

[92]   Lin, supra note 83, at 1332.

[93]   See Laura Cruz-Castro and Luis Sanz-Menendez, "Politics and Institutions: European Parliamentary Technology Assessment," 72 TECH. FORECAST. SOC. CHANGE 429, 432–5 (2004); Van Eijndhoven, supra note 82, at 271–4.

[94]   Lin, supra note 83, at 1332–4.

[95]   By formal methods, I refer to data collection and evaluation methodologies for the range of scientific, economic and social issues addressed by the assessment, discussed below. See Patrick Hofstetter, "Tools for Comparative Analysis of Alternatives: Competing or Complementary Perspectives," 22 RISK ANALYSIS 833 (2002) (discussing a variety of tools used to evaluate the health, economic and social impacts of technology).

basis.[96] Over time the available methods for technology assessment have evolved, growing more sophisticated in range of factors considered, methods for dealing with uncertainty, and the evaluation of tradeoffs. Examples include consequential life cycle assessment,[97] sustainability analysis,[98] alternatives analysis,[99] technology options analysis,[100] tradeoff analysis,[101] and benefit cost analysis.[102] These tools address different levels of analysis (for example, product/service; technology/industry) and types of impacts (ecological, health, economic) through a variety of methodological approaches. Their uses and limitations, and the challenges of integrating them, are beyond the scope of this chapter. They all share the basic goal, however, of identifying and evaluating the relevant impacts of technology. The question to which I now turn is whether such assessments would assist policy makers in avoiding unintended consequences of technological changes that are encouraged or mandated by their policies. In answering that question, it is helpful to return to the three categories of factors that contribute to unintended consequences: attention/information processing barriers, inherent unforeseeability, and political influence.

Formal technology assessment could offer significant help in overcoming many of the attention and information barriers identified in Section 6.3, above. The first two barriers involve situations in which relevant information is lacking for substantive reasons; that is, either it is too costly to obtain (rational ignorance) or it is simply not available

---

[96] Fred B. Wood, "Lessons in Technology Assessment Methodology and Management at OTA," 54 TECH. FORECAST. SOC. CHANGE 145, 153–4 (1997).

[97] D. Rajagopal, G. Hochman and D. Zilberman, "Indirect Fuel Use Change (IFUC) and the Lifecycle Environmental Impact of Biofuel Policies," 39 ENERGY POL'Y 228 (2011).

[98] Clara Rosalía Álvarez-Chávez, Sally Edwards, Rafael Moure-Erasoa and Kenneth Geiser, "Sustainability of Bio-Based Plastics: General Comparative Analysis and Recommendations For Improvement," 23 J. CLEANER PROD. 47 (2012).

[99] Timothy F. Malloy, Peter Sinsheimer, Ann Blake and Igor Linkov, "Use Of Multi-Criteria Decision Analysis In Regulatory Alternatives Analysis: A Case Study Of Lead Free Solder" INTEGRATED ENVIRONMENTAL ASSESS-MENT AND MANAGEMENT (forthcoming).

[100] Nicholas A. Ashford and Ralph P. Hall, TECHNOLOGY, GLOBAL-IZATION, AND SUSTAINABLE DEVELOPMENT: TRANSFORMING THE INDUSTRIAL STATE 392–5 (2011).

[101] Id. at 396–400; Hofstetter et al., supra note 95, at 839–40; Graham and Wiener, supra note 43.

[102] Hofstetter et al., supra note 95, at 841–2.

(uncertain/incomplete information). Regarding rational ignorance, "value of information" (VoI) techniques can assist policy makers in determining whether particular data is worth pursuing, in terms of cost and time. Traditional VoI analysis "identifies uncertainties whose resolution has a good chance of changing the decision,"[103] allowing for informed judgment about whether additional cost or delay incurred in further research is justified.[104] Technology assessment can also assist policy makers in making decisions under conditions of uncertainty, whether salient information is simply unavailable, or too costly to obtain. Here sophisticated methods, such as Monte Carlo risk analysis and sensitivity analysis, can alternately delineate a range of potential outcomes,[105] or specify the sensitivity of the assessment to changes in the uncertain data.[106]

The attention and information barriers of asymmetric information, omitted voices and fragmentation arise from structural aspects of the decision making environment. Information-flow to the legislative decision maker is blocked because the party with the information withholds it, is excluded from the legislative process, or is unable to free the information from its own organization. Depending on its particular institutional design, technology assessment could mitigate these barriers by creating a procedural framework that encourages or even mandates the generation and submission of information to the legislature. A structured systematic data collection method, coupled with sufficient authority to collect information, would be more likely to unearth withheld data, identify stakeholders with relevant information, and even overcome fragmentation across multiple regulatory agencies or offices. Consider the case of MTBE, in which the oil industry had generated significant information regarding potential problems regarding toxicity and mobility of the compound in groundwater. If sufficient information collection

---

[103]   Igor Linkov, Matthew E. Bates, Benjamin D. Trump, Thomas P. Seager, Mark A. Chappell and Jeffrey M. Keisler, "For Nanotechnology Decisions, Use Decision Analysis," NANO TODAY (forthcoming).

[104]   See Igor Linkov, Matthew E. Bates, Laure J. Canis, Thomas P. Seager and Jeffrey M. Keisler, "A Decision-Directed Approach or Prioritizing Research into the Impact of Nanomaterials on the Environment and Human Health," 6 NAT. NANOTECHN. 784 (2011).

[105]   Alan M. Porter, Frederick A. Rossini and Stanley R. Carpenter, A GUIDEBOOK FOR TECHNOLOGY ASSESSMENT AND IMPACT ANALYSIS (1980).

[106]   Simon French and Jutta Geldermann, "The Varied Contexts of Environmental Decision Problems and their Implications for Decision Support," 8 ENVTL. SCI. & POL'Y 378, 387–8 (2005) (discussing sensitivity analysis).

authority had been given to an agency like the Office of Technology Assessment, such an agency might have extracted that information.

The impact of technology assessment on the last attention/information barrier, heuristics and cognitive biases, is less clear. Some such biases affect the search for relevant information (leading individuals to overlook relevant data) or the ability to judge the likelihood that events may occur.[107] For these, formal technology assessment may effectively overcome the bias,[108] assuming that the assessment adequately delineates the necessary data and clearly presents the relevant probabilities.[109] Other biases affect how the individual evaluates the available information; for example, how much weight the individual accords the facts and conclusions presented. Where an individual has a pre-existing, strong preference in terms of ultimate outcome, pre-decisional biases may affect the decision making process so as to advance that preferred outcome.[110] In such cases, "motivated reasoning" may occur, in which legislators subconsciously slant their evaluation of information and recommendations in favor of a politically desired result.[111] The availability of technology assessment in and of itself is unlikely to mitigate these pre-decisional biases. After all, the very essence of motivated reasoning is the skewed evaluation of unwelcome information.[112]

---

[107] Gary Mucciaroni and Paul J. Quirk, DELIBERATIVE CHOICES 28–9 (2006); Jeffrey J. Rachinski, "Heuristics, Biases, and Governance," in BLACKWELL HANDBOOK OF JUDGMENT AND DECISION MAKING 567, 575–6 (Derek J. Koehler and Nigel Harvey, eds, 2004).

[108] Jeffrey J. Rachlinski and Cynthia R. Farina, "Cognitive Psychology and Optimal Government Design," 87 CORNELL L. REV. 549, 574–5 (2001–2002).

[109] Paul Slovic, Bruce Fischhoff and Sarah Lichtenstein, "Perceived Risk: Psychological Factors and Social Implications," 376 PROC. R. SOC. A 17, 29–31 (1981).

[110] Timothy F. Malloy, "Disclosure Stories," 32 FLA. ST. U. L. REV. 617, 650–3 (2004–2005).

[111] Id.; Ziva Kunda, "The Case for Motivated Reasoning," 108 PSYCHOL. BULL. 480 (1990).

[112] Increased accountability – defined as the expectation that one will have to justify one's views to others – may play some role in "de-biasing" the decision process. Malloy, Disclosure Stories, supra n. 109, at 663–4; Jennifer S. Lerner and Philip E. Tetlock, "Accounting for the Effects of Accountability," 125 PSYCHOL. BULL. 255, 255–6 (1999); Philip E. Tetlock, "The Impact of Accountability on Judgment and Choice: Toward a Social Contingency Model," in 25 ADVANCES IN EXPERIMENTAL SOC. PSYCHOL. 331, 337 (Mark P. Zanna, ed., 1992). Such an expectation is likely to be met in the form of debates

Technology assessment may also mitigate inherent unforeseeability, the second category of factors contributing to unintended consequences. This category acknowledges that some systems, be they natural such as an ecosystem or social such as a complex market, are so complex as to defy prediction methods that use narrative scenario-analysis and conventional probabilistic modeling.[113] These are typically situations of deep uncertainty regarding either the operation of the system in question, or the relevant data, such as climate change or military strategy.[114] Technology assessment in such circumstances involves new forms of decision analysis that make use of expanding computational capacities, including agent-based modeling and robust decision analysis.[115]

The last category, political influence, is perhaps the most problematic in terms of the impact of technology assessment. The key issue here is whether the ostensibly objective information and evaluation provided by technology assessment can overcome or at least moderate the ideological constraints[116] and political motivations that can so strongly influence legislators' decisions. The literature on the use of policy analysis by Congress provides some general insights regarding this question. Rarely does policy analysis submitted to legislators give rise to a specific policy outcome.[117] Rather, legislators, their staff and committees tend to use policy analysis in three major ways: substantive, elaborative and strategic. Substantive use occurs where the legislator relies on the analysis for factual background and to identify a set of relevant issues for consideration.[118] Elaborative use refers to using the analysis to fill in specific

---

within Congressional committees and on the floor. That said, any de-biasing effect may be overshadowed by the role of political influence (discussed below).

[113]   Robert J. Lempert, "A New Decision Sciences for Complex Systems," 99 PROC. NAT'L ACAD, SCI. 7309 (2002).

[114]   Id. at 7310.

[115]   Id. at 7310–13; Robert J. Lempert and Myles T. Collins, "Managing the Risk of Uncertain Threshold Responses: Comparison of Robust, Optimum, and Precautionary Approaches," 27 RISK ANALYSIS 1009 (2007).

[116]   For these purposes, ideology refers to "abstract beliefs about the proper role of government." Frances E. Lee, "Interests, Constituencies and Policy Making," in THE LEGISLATIVE BRANCH 281, 298 (Paul J. Quirk, Paul J. Quirk and Sarah A.Binder, eds, 2005).

[117]   Nancy Shulock, "The Paradox of Policy Analysis: If It Is Not Used, Why Do We Produce So Much of It?," 18 J. POL. ANALYSIS & MANAG. 226, 227–8 (1999); Carol H. Weiss, "Congressional Committees as Users of Analysis," 8 J. POL'Y ANALYSIS & MGMT. 411, 428 (1989).

[118]   David Whiteman, "The Fate of Policy Analysis in Congressional Decision Making: Three Types of Use in Committees," 38 W. POL. QUARTERLY 294,

details of a policy and to check the policy for major gaps or flaws.[119] Policy analysis is used strategically when the legislator relies upon it to support an existing position or undercut a competing policy proposal. In strategic use, the user will typically ignore or challenge analysis that is contrary to their preferred position.[120]

Whiteman surveyed the use of five OTA projects by Congress in the 1980s, concluding that all three types of use were present, with substantive use being the least prevalent. Elaborative use was quite common, particularly with respect to information gathering activities. Strategic use was the most prevalent, and its dominance was substantially magnified in cases involving high levels of conflict within Congress.[121] Weiss obtained consistent results in her later study of policy analysis, also concluding that strategic use was the most common.[122] Thus, the available information regarding congressional use of policy analysis suggests that technology assessment could play a meaningful role in some aspects of congressional deliberation. In politically charged environments, however, it appears less likely that technology assessment will be able to overcome policy positions rooted in political motivations.

6.6 CONCLUSION

We end where we started, with the premise that if government promotes new technologies, it also should minimize their significant unintended consequences. Legislative technology assessment can be an important tool in that effort. It has the potential to mitigate many factors that contribute to unintended consequences. However, we must be realistic about its limits. Most notably, the overtly political nature of legislative decision making can undermine the effectiveness of technology assessment, transforming it into an instrument of strategic action rather than an aid to careful deliberation.

298–9 (1985). Weiss essentially calls this use "guidance." Weiss, supra note 117, at 424.

[119] Whiteman, supra note 118, at 299.
[120] Id.; Weiss, supra note 117, at 424.
[121] Whiteman, supra note 118, at 299–304.
[122] Weiss, supra note 117, at 424.

7. Governing the governance of emerging technologies

Gary E. Marchant[1] and Wendell Wallach

7.1 INTRODUCTION

Emerging technologies such as nanotechnology, biotechnology, personalized medicine, synthetic biology, applied neuroscience, geoengineering, social media, surveillance technologies, regenerative medicine, robotics and artificial intelligence present complex governance and oversight challenges. These technologies are characterized by a rapid pace of development, a multitude of applications, manifestations and actors, pervasive uncertainties about risks, benefits and future directions, and demands for oversight ranging from potential health and environmental risks to broader social and ethical concerns.[2] Given this complexity, no single regulatory agency, or even group of agencies, can regulate any of these emerging technologies effectively and comprehensively.

What has begun to emerge, and what is inevitable given the complex dynamics of emerging technologies, is a matrix of different government regulatory agencies, statutes, regulations and guidances, supplemented by an even more tangled web of "soft law" approaches such as codes of conduct, statements of principles, partnership programs, voluntary programs and standards, certification programs and private industry initiatives. That is just at the national level – but because these emerging technologies tend to be international in scope and application, the many

[1] The research and writing of this chapter was supported in part by a National Science Foundation grant (Award SES-0921806) to GEM.
[2] Gary E. Marchant, "Conclusion: Addressing the Pacing Problem," in Gary Marchant, Braden Allenby and Joseph Herkert (eds), THE GROWING GAP BETWEEN EMERGING TECHNOLOGIES AND LEGAL-ETHICAL OVERSIGHT: THE PACING PROBLEM 199, 199 (2011); Gregory N. Mandel, "Regulating Emerging Technologies," 1 LAW, INNOVATION TECH. 75, 75–7 (2009).

different national and international regulatory and non-regulatory instruments make the oversight landscape even more complicated.

While the diversity of oversight actors, mechanisms, approaches and requirements for any emerging technology are unavoidable at least at this time, it does create problems with regard to coherence, potential overlaps and gaps, and coordination. We therefore propose the creation of emerging technology coordination committees, which would be public–private consortia that do not directly regulate, but rather serve a coordinating function to identify, publicize, evaluate for gaps, overlaps and conflicts, and coordinate the various governance instruments and proposals that exist for a given emerging technology. In short, each of these emerging technologies needs an issue manager, the role that the proposed coordinating committees are intended to fulfill. Of course, there will be many challenges and potential problems with such a novel approach. For example, the effectiveness of any new body in coordinating the activities of various stakeholders will depend upon those parties perceiving cooperation as being in their interest. In addition, a coordinating body will only be effective if it has some economic or political influence. Notwithstanding the acknowledged limitations and challenges of our proposal, the urgent need for a coordinating function for emerging technologies justifies exploration of the feasibility and design of such coordinating committees.

In this chapter, we first briefly summarize the dynamics and complexity of emerging technologies that create the need for coordinating committees. Next, we identify some previous attempts to create institutions to manage an emerging technology, describing their successes but also where they fall short. In many respects, these prior endeavors were designed as de facto attempts to fill the coordination gap we seek to address here, so much can be learned from these prior experiments. Finally, we propose the creation of coordinating committees, including preliminary thoughts on their structure, function and implementation.

7.2 THE GOVERNANCE CHALLENGE OF EMERGING TECHNOLOGIES

There are many emerging technologies today, ranging from nanotechnology to geoengineering, synthetic biology to robotics, and applied neuroscience to surveillance technologies. These diverse technologies differ in their details, focus and in other important ways – yet share a common set of attributes that makes their regulation difficult. First, these emerging technologies are all developing at a very fast pace, in many cases on exponential or near-exponential paths. At the rapid rate of change, emerging technologies leave behind traditional governmental regulatory models and approaches which are plodding along slower today than ever before.[3]

Emerging technologies are also characterized by high uncertainties about their risks, benefits and future direction. Additional complexity is provided by the large range and diversity of types, applications and participants within each emerging technology category, not limited as were previous technologies to a single industry sector or application. Finally, the types of concerns raised by emerging technologies go beyond the health and environmental risks traditionally covered by regulatory statutes to also include broader ethical, social and economic concerns, including privacy, fairness and human enhancement issues.[4]

Matched against these intrinsic characteristics of emerging technologies is the growing dysfunction of our traditional regulatory systems.[5] Governments are heavily involved in directing scientific development in the form of capital investment and through regulations and regulatory oversight directed at public health, protecting the environment, protecting human subjects and animals used in research, the safety of goods and services, the safety of laboratory workers, and, more recently, biosecurity. Yet, as technology speeds up, our regulatory system is increasingly ossified, slowing down and becoming more and more burdened by bureaucratic red tape.[6] Regulatory agencies are forced to operate under outdated statutes often written 30 or 40 years ago, statutes that did not anticipate and do not fit the emerging technologies they are now being called upon to regulate. The hope for new, updated legislative authority is bleak short of some disaster that will spark Congress into action. Moreover, there is increasing political resistance to regulations, which are perceived as interfering with economic productivity through inflexible, shortsighted, and overly bureaucratic constraints. In addition,

---

[3]   Gary E. Marchant, "The Growing Gap Between Emerging Technologies and the Law," in Gary Marchant, Braden Allenby and Joseph Herkert (eds), THE GROWING GAP BETWEEN EMERGING TECHNOLOGIES AND LEGAL-ETHICAL OVERSIGHT: THE PACING PROBLEM 19, 19–33 (2011).

[4]   Gary Marchant, Ann Meyer and Megan Scanlon, "Integrating Social and Ethical Concerns Into Regulatory Decision-Making for Emerging Technologies," 11 MINNESOTA J. LAW SCIENCE & TECHNOLOGY 345, 345–63 (2010); Albert C. Lin, "Revamping Our Approach to Emerging Technologies," 76 BROOKLYN L. REV. 1309 (2011).

[5]   Marchant, supra note 3, at 22–5.

[6]   Thomas O. McGarity, "Some Thoughts on Deossifying the Rulemaking Process," 41 DUKE L.J. 1385, 1385–1462 (1992).

the funding for regulatory oversight is insufficient and in many cases declining due to overall government budgetary pressures, creating merely the illusion that the public is being protected from various harms.

The combination of these complex emerging technologies with an increasingly outdated and obsolete regulatory system means that traditional regulatory approaches will provide insufficient oversight of emerging technologies. For example, consider the case of nanotechnology, where governments are struggling and mostly failing to provide meaningful regulation. The problems are numerous.[7] The risks of nanotechnology cannot be accurately assessed, as scientists are only in the early stages of understanding the potential hazards of nanomaterials and the complex set of factors influencing hazard potential.[8] No standardized toxicology assays are available, nor is quantified risk assessment (required by many regulatory statutes) a possibility anytime soon.[9] The enormous range of types, processes and applications of nanotechnology also prevents a consistent and comprehensive regulatory approach.[10] In fact, the diversity of nanomaterials is challenging regulators from even defining nanotechnology in a scientifically credible and legally viable way.[11] The rapid evolution of nanotechnology creates a moving target that regulation will always lag at least one generation behind.[12] Finally, while regulators struggle to address the safety and environmental risks of nanotechnology, they are even less equipped or not even authorized to address the more complex risks such as privacy and human enhancement that will be presented by the next generation of nanotechnology products.[13]

---

[7] Gary E. Marchant et al., "The Biggest Issues for the Smallest Stuff: Nanotechnology Regulation and Risk Management," 52 JURIMETRICS 243, 256–65 (2012).

[8] Stephan T. Stern and Scott E. McNeil, "Nanotechnology Safety Concerns Revisited," 101 TOXICOLOGICAL SCI. 4, 16 (2008).

[9] Katherine A. Clark, Ronald H. White and Ellen K. Silbergeld, "Predictive Models for Nanotoxicology: Current Challenges and Future Opportunities," 59 REG. TOXICOLOGY PHARMACOLOGY 361, 361 (2011).

[10] US Environmental Protection Agency, NANOTECHNOLOGY WHITE PAPER 29 (February 2007), available at http://www.epa.gov/osa/pdfs/nanotech/epa- nanotechnology-whitepaper-0207.pdf (accessed July 12, 2013).

[11] Andrew D. Maynard, "Don't Define Nanomaterials," NATURE, July 27, 2011, at 31, 31; Marchant et al., supra note 7, at 260–1.

[12] David Rejeski, "The Next Small Thing," ENVTL. F, Mar.–Apr. 2004, at 45, 45.

[13] J. Clarence Davies, OVERSIGHT OF NEXT GENERATION NANO-TECHNOLOGY (Project on Emerging Nanotechnologies Report 18, April 2009).

The trials and tribulations of nanotechnology regulation are not unique – most emerging technologies are facing similar regulatory roadblocks, if not worse. Moreover, even if it were possible to create a credible and effective regulatory regime at the time of adoption, it would not remain so for long given how fast these emerging technologies are evolving. Thus new regulations are commonly responses to yesterday's challenge. They seldom institute mechanisms that are responsive to change and the idiosyncratic ways in which technological possibilities unfold and an industry develops over time. These risks of premature or outdated regulations further discourage regulatory action, coalescing with fears that over-regulation may burden technological innovation and stifle economic growth. In light of the resistance to over-regulation, new regulations are often postponed until there is a disaster whose repetition must be avoided.

Regulatory agencies are also often and increasingly limited by resource constraints. The administrators of the agency are concerned that they will be held responsible for new harms which might occur under their watch, but they are seldom given adequate funding for staff and the other resources needed to adequately fulfill new or even existing mandates. So the FDA, for example, is dependent on the sporadic reporting of adverse incidents by hospitals, providers, public health officials, research institutions and patients, and is only capable of investigating fully the most serious of those incidents.[14]

Many other stakeholders have an interest and responsibility in the safe development of emerging technologies, including industry, the public, scholars, non-governmental organizations, and the media. Industry has the most to gain or lose from oversight systems. Leaders of industry are particularly concerned with limiting the imposition of regulations that they perceive as hampering the development of goods and services in irrational ways. However, responsible leaders of industry are also cognizant that they will be held responsible for both the fact and perception that their products and services cause harm. Self-regulation by industry is certainly one option, but stimulating successful vehicles for self-regulation will demand that industry leaders perceive such mechanisms as in the self-interest of their companies and agree to comply with such measures voluntarily.

---

[14]   Diane K. Wysowski and Lynette Swartz, "Adverse Drug Event Surveillance and Drug Withdrawals in the United States, 1969–2002: The Importance of Reporting Suspected Reactions," 165 Arch. Internal Med. 1363, 1363–9 (2005).

The public generally perceives technology as an engine of innovation, productivity and novelty, but is also concerned with the minimization of harms. The scholarly community is a source of innovative ideas, a critic of irrational or unsubstantiated claims, and a watchdog underscoring societal, ethical, and policy challenges that have not been addressed. Scientific research performed within the academy is a valuable tool for determining whether ideas or claims made by industry are grounded in evidence. Social theorists can be helpful in testing whether specific governmental policies are actually addressing the challenge for which the policy was designed. Like academics, NGOs commonly focus upon specific concerns and serve as advocates for issues they believe need to be addressed. But both academics and NGOs feel that their evidence, issues, and warnings all too often go either unnoticed or ignored.

The media, at its best, serves as a disseminator of essential information and ideas, packaged with a little education. Policy debates are commonly worked through under the spotlight of media coverage. But, at its worse, the media can sensationalize and exacerbate bias, rumors, falsehood, and unwarranted fears. When media outlets do try to be fair and present both or all sides of an issue, they commonly contribute to the illusion that competing positions are of equal stature, which is often not the case.

Monitoring and managing the development of emerging technologies requires more than just government regulation, given its limitations, and must expand to a "governance" model.[15] Governance expands the circle of responsibility for managing a problem beyond government regulators to also include other entities including industry, non-governmental organizations, insurers, think tanks, courts and other national and international entities.[16] While government regulation still plays an essential role, other entities need to supplement government regulations with various types of "soft law" oversight mechanisms including codes of conduct, public-private partnership programs, voluntary programs and standards, guidelines, certification schemes, and various types of other industry initiatives.[17]

---

[15] O. Renn and M.C. Roco, "Nanotechnology and the Need for Risk Governance," 8 J. NANOPARTICLE RES. 153 (2006).

[16] Kenneth W. Abbott and Duncan Snidal, "Strengthening International Regulation Through Transnational New Governance: Overcoming the Orchestration Deficit," 42 VAND. J. TRANSNAT. L. 501, 506–9 (2009).

[17] Diana Bowman and Graeme A. Hodge, "A Big Regulatory Tool-Box for a Small Technology," 2 NANOETHICS 193, 198–202 (2008); Kenneth W. Abbott & Duncan Snidal, "Hard and Soft Law in International Governance," 54 INT'L ORG. 421, 434–50 (2000).

These "soft law" approaches have many potential benefits including that they tend to be participatory, cooperative, reflexive, adaptive, involve a mixture of tools, and invoke resources and responsibility at multiple levels and from multiple parties.[18] Of course, these soft law approaches also have their weaknesses, most prominently that they are generally non-enforceable and do not offer the procedural opportunities and protections that are provided by traditional government regulation.[19] Although soft law approaches have clear weaknesses, they are an inevitable and integral component of the governance of emerging technologies given the limitations of traditional regulatory approaches.

One previously under-recognized problem with soft law approaches is a coordination problem. There are usually many soft law initiatives, not just a single one, for any emerging technology. For example, there are at least 11 major soft law initiatives in place for nanotechnology oversight. This proliferation of different soft law programs and proposals is a function of one of the strengths of soft law approaches – governance initiatives are not limited to centralized federal agencies, but rather can be launched by any entity or coalition of organizations. What is important but lacking is a credible approach for facilitating the coordination of the independent concerns and actions of these entities, including reconciling these soft law initiatives not only with each other but also with traditional government regulation.

7.3 NEED FOR AN "ISSUE MANAGER"

To date, governance of emerging technologies have generally proceeded with a piecemeal approach, in which government agencies and developers of soft law programs propose new oversight initiatives one piece at a time, with little regard to other initiatives affecting the same technology. There are, however, serious doubts as to whether the problematic challenges posed by the rapid and unpredictable emergence of new technological possibilities can be effectively managed by this piecemeal,

---

[18]   Abbott and Snidal, supra note 16, at 554–58; Kenneth Abbott, Elizabeth Curley and Gary Marchant, "Soft Law Oversight Mechanisms for Nanotechnology," 52 JURIMETRICS 279, 300–7 (2012).

[19]   Steffen Foss Hansen and Joel A. Tickner, "The Challenges of Adopting Voluntary Health, Safety and Environment Measures for Manufactured Nanomaterials: Lessons from the Past for More Effective Adoption in the Future," 4 NANOTECH. LAW & BUSINESS 341 (2007); Timothy Malloy, "Soft Law and Nanotechnology: A Functional Perspective," 52 JURIMETRICS 347, 355 (2012).

incremental approach to monitoring and managing their development. The emerging technologies will likely require a more coordinated, holistic and nimble approach, while not sacrificing diligence in overseeing discernible dangers. In short, emerging technologies need an "issue manager" – some entity to coordinate and serve as the central hub for the governance of that technology. An issue manager might be understood as a committee that provides a focus, a field of play, and a voice for bringing together the different approaches, perspectives, and functions relating to the governance of a given technology.

This institutional issue manager would not be a regulator itself, but would act more like an orchestra conductor in trying to harmonize and integrate the various governance approaches that have been implemented or proposed by others.[20] Its functions might include collecting and reporting on information about existing governance programs for an emerging technology, coordinating actors by identifying gaps, overlaps and inconsistencies with existing and proposed programs, providing a forum for stakeholders to deliberate on governance of a technology, producing recommendations, reports and roadmaps on the governance of a technology, and serving as a trusted "go to" source for stakeholders, the media and the public for information about the technology and its governance.

There are a number of precedents, models and proposals for such an "issue manager" for an emerging technology. Some are summarized below, along with their limitations:

- *Coordinated Framework for Biotechnology:* in 1986, the White House Office of Science and Technology Policy published the Coordinated Framework for Regulation of Biotechnology Products, a matrix of statutes and agencies to regulate the variety of products manufactured by biotechnology.[21] This was the first time the federal government had published a comprehensive map of the regulation of an emerging technology that was expected to produce a variety of products. Different classes of products were placed under the jurisdiction of different federal agencies and statutes. While a

---

[20] See Kenneth W. Abbott and Duncan Snidal, "International Regulation Without International Government: Improving IO Performance Through Orchestration," 5 REV. INT'L ORGS. 315, 318–23 (2010) (developing concept of orchestration of governance actors).

[21] Coordinated Framework for Regulation of Biotechnology, 51 Fed. Reg. 23,302 (June 26, 1986).

useful and effective starting point for the oversight of biotechnology, the Coordinated Framework had several limitations. It only included government programs and not any voluntary or other non-governmental oversight, had several gaps and overlaps in its coverage of governmental regulatory programs, provided no ongoing structure for stakeholder involvement and deliberation, and has never been updated.[22]

● *National Nanotechnology Initiative (NNI)*: similar in many ways to the Coordinated Framework for Biotechnology, the NNI functions as a coordinating body for the federal government's nanotechnology initiatives. The NNI is administered through a small National Nanotechnology Coordination Office that supports an inter-agency Nanoscale Science, Engineering and Technology (NSET) Subcommittee, consisting of representatives of 26 federal agencies. The NSET has established four Working Groups which define many of the activities of the NNI: (i) the Global Issues in Nanotechnology Working Group; (ii) the Nanotechnology Environmental & Health Implications Working Group; (iii) the Nanomanufacturing, Industry Liaison & Innovation Working Group; and (iv) the Nanotechnology Public Engagement & Communications Working Group. The NNI is probably the model that comes closest to the coordinating committee model we set forth in this chapter, with the major difference being that it is solely governmental. While the NNI reaches out to various stakeholders in the nanotechnology arena, they are not actual members of the NNI, which is restricted to governmental entities and personnel.

● *ICON*: the International Council on Nanotechnology (ICON) is perhaps the closest real-world attempt to implement the type of coordinating function we propose here for technology coordinating committees. ICON is an international multi-stakeholder organization with a mission to "develop and communicate information regarding potential environmental and health risks of nanotechnology."[23] ICON's activities included sponsoring multi-stakeholder forums to advance the understanding of health and

---

[22] Gregory N. Mandel, "Gaps, Inexperience, Inconsistencies, and Overlaps: Crisis in the Regulation of Genetically Modified Plants and Animals," 145 WM. & MARY L. REV. 2167 (2003–2004).

[23] ICON, FACTS ABOUT THE INTERNATIONAL COUNCIL OF NANOTECHNOLOGY (rev. October 2009), available at http://cohesion.rice.edu/centers andinst/icon/emplibrary/ICON%20Fact%20Sheet_102609.pdf (accessed July 12, 2013).

environmental risks of nanotechnology, creating an online database and virtual journal of published studies on the risks of nano-materials, and creating an online wiki site of best practices for handling nanomaterials, called the GoodNanoGuide.[24] ICON's main source of funding was a $50,000 per year membership fee from industry members, which caused concern and reluctance to participate by some NGOs.[25] The main difference between ICON and the coordinating committees proposed here is that ICON focused primarily on the scientific and risk management aspects of the technology, whereas the coordinating committees would give greater emphasis to the governmental and non-governmental over-sight and governance mechanisms that are in place or have been proposed.

- *Pew Centers:* the Pew Charitable Trusts have created several initiatives that have served as a *de facto* issue manager for specific emerging technologies, such as the Pew Initiative on Food and Biotechnology (2001–2007), the Project on Emerging Nanotech-nologies (2005–2011) and the Pew Center on Global Climate Change (1998–2011). These projects became focal points in the governance of the technology they each individually addressed by issuing influential reports, identifying key gaps and needs in regulatory programs, and serving as a forum for stakeholders to deliberate about the technology. For example, the Pew Initiative on Food and Biotechnology produced over 20 influential reports, fact sheets and briefings on various regulatory, social, economic, polit-ical and ethical implications of biotechnology products, produced a critical analysis of the US Coordinated Framework, and convened a Stakeholder Forum that assembled representatives from industry, academia, consumer and environmental groups to find consensus on recommendations to enhance US regulatory oversight of biotech-nology products. Although no consensus was ultimately reached, the initiative did bring the various interests together for productive dialogue and a better understanding of the issues. Based on these accomplishments, an editorial in *Nature Biotechnology* described the Pew Initiative as "the go-to resource for neutral and trustworthy

---

[24] .Kristen M. Kulinowski and Matthew P. Jaffe, "The GoodNanoGuide: A Novel Approach for Developing Good Practices for Handling Engineered Nano-materials in an Occupational Setting," 6 NANOTECHNOLOGY L. & BUS. 37 (2009).

[25] ICON, supra note 22; Abbott et al., supra note 18, at 295–6.

information on agbiotech."[26] Unfortunately, the Pew Foundation ceased funding of such initiatives in 2011.

● *White House Emerging Technologies Interagency Policy Coordination Committee*: the White House Office of Science and Technology Policy (OSTP), Office of Information and Regulatory Affairs (OIRA), and the US Trade Representative (USTR) created the White House Emerging Technologies Interagency Policy Coordination Committee in 2010. The committee includes assistant secretary-level representatives from about 20 federal agencies. Its goal is "open consideration of policy questions" raised by emerging technologies "with the full range of stakeholders, including governments, industry, non-governmental organizations, academia, and the public."[27] Although the committee published in its first year a list of principles for the governance of emerging technologies, it has maintained a low profile (if not completely dormant) since releasing this initial document, and has provided no public forum for stakeholders to interact or participate.

These five examples (and there are no doubt many others that could be cited) indicate an interest and demand for some type of coordinating body for governing emerging technologies. While the examples summarized above have all made substantial contributions to the governance of specific emerging technologies, they are each limited in different ways in their scope, activities, visibility, participation and longevity. A more dynamic, inclusive, high profile and stable entity is needed to fill the gap in the governance of emerging technologies.

In addition to the existing examples described above, there have been occasional proposals for some type of coordinating institution for various emerging technologies. For example, Daniel Fiorino, in his report on the numerous voluntary programs that have sprung up to help manage nanotechnology, proposed the creation of a "Nano Stewardship Council" that he described as follows:

---

[26] Editorial, "Hearts and Minds," 25 NATURE BIOTECHNOLOGY 143 (2007).

[27] John P. Holdren, Cass R. Sunstein and Islam A. Siddiqui, MEMORANDUM FOR THE HEADS OF EXECUTIVE DEPARTMENTS AND AGENCIES: PRINCIPLES FOR REGULATION AND OVERSIGHT OF EMERGING TECHNOLOGIES (March 11, 2011), available at http://www.whitehouse.gov/sites/default/files/omb/inforeg/for-agencies/Principles-for-Regulation-and-Oversight-of-Emerging-Technologies-new.pdf (accessed July 12, 2013).

NGOs, business, government and others should establish a Nano Policy Forum for discussing nano oversight issues and developing needed tools. This body should be tasked with considering the long-term value and development of a multi-stakeholder Nano Stewardship Council modeled generally on other collaborative mechanisms. The forum could be funded in equal amounts by government, foundation and business resources or be the subject of a specific congressional appropriation. It would provide an ongoing, neutral forum for discussions on nanotechnology policy issues and options and a clearinghouse for information.[28]

Similarly, the Presidential Commission for the Study of Bioethical Issues called for the creation of a "coordinating body" within the government for synthetic biology that, among other things, would: (1) leverage existing resources by providing ongoing and coordinated review of developments in synthetic biology, (2) ensure that regulatory requirements are consistent and non-contradictory, and (3) periodically and on a timely basis inform the public of its findings."[29]

7.4 PROPOSAL FOR GOVERNANCE COORDINATING COMMITTEES

A governance coordinating committee (GCC) for individual emerging technologies offers the advantage of coordinating the development and management of that technology comprehensively. A nanotechnology coordinating committee, for example, would be charged with facilitating the development and integration of both governmental and "soft law" mechanisms for promoting the safe use of nanotechnology, including development and coordination of test methods, best practices, and risk management methods. The GCC would not replicate the activities of other bodies such as research institutions or regulatory authorities by developing its own standards, definitions and policies, but would rather seek to identify gaps, overlaps, inconsistencies and synergies in the existing public and private enactments. It would also engage key stakeholders in cross-communication and consensus-building efforts, and serve as a focal point for the media and public on nanotechnology

---

28 Daniel J. Fiorino, VOLUNTARY INITIATIVES, REGULATION, AND NANOTECHNOLOGY OVERSIGHT: CHARTING A PATH 42–3 (Project on Emerging Nanotechnologies Report 19, November 2010).
29 Presidential Commission For The Study Of Bioethical Issues New Directions, THE ETHICS OF SYNTHETIC BIOLOGY AND EMERGING TECHNOLOGIES 8 (December 2010).

governance issues. The key challenge for a GCC is to forge a robust set of mechanisms that address the challenge comprehensively while being flexible enough to change as the industry develops. In forging this robust system of mechanisms, the charge of the GCC will be to stay lean and nimble, to emphasize inclusiveness, and to utilize existing resources to the maximum extent possible.

The influence and effectiveness of a GCC will depend on its ability to help the various stakeholders appreciate why cooperation in the building of a robust set of policy mechanisms lies in their interest. For the industry it oversees, as well as other stakeholders for that technology, a GCC will need to establish itself as a good-faith broker. If the public, legislature, executive branch, and private stakeholders perceive the GCC as a credible authority on the governance of the applicable technology, then its recommendations will hold some weight. It can provide industry with a roadmap and recommendations for acceptable risk management practices and the responsible governance of the technology. It may also be able to use the carrot of recommending policies that have the effect of lowering or limiting liability risks or regulatory burden. In many new fields companies will not move forward in the development of new technologies, even those that provide broad benefits with low or uncertain risks, until their responsibility for those risks can be established. For example, if the robotics industry establishes rigorous procedures for certifying the use of specific service robots for specific tasks, then a mechanism such as no-fault insurance for harms that could not be foreseen might be recommended.

The executive and legislative branches of government are likely to take seriously the recommendations of the GCCs if they consider these committees to be credible and diligent. Presumably recommendations by the GCC to the legislature or executive branch agencies will be made only when the GCC has determined that there is a serious gap in the existing mechanisms for monitoring and managing risks, and once other options are exhausted. The GCC will need to be vigilant in not usurping the authority of regulatory agencies. We believe that the regulatory authorities will view working together with a credible GCC as in their interest. Most existing agencies are overwhelmed by unfunded mandates and the fear that they will be held responsible when harms within their charter occur, and so should welcome the buffer and at least interim coverage that a GCC can provide while agencies develop the data, test methods, and regulatory findings they need to take regulatory action.

In forging a comprehensive understanding of the challenges posed by a new technological development, a GCC will need to be exposed to research done by scientists and social scientists as well as listen to concerns raised by social critics and NGOs. These communities will not necessarily feel that their perspectives are always given adequate weight, but will appreciate that someone is factoring these considerations into a more comprehensive understanding of the challenges and providing a meaningful forum where their concerns will be heard and considered. In turn, the GCC will be reporting to funding sources such as the NSF, NIH, DARPA, and foundations regarding research that needs to be performed in order for society to better manage the potential risks and ethical and social challenges arising from new technologies.

Moreover, the GCC may be helpful in modulating the rate of development of an emerging technology. Just as an economy can stagnate or overheat, so also can technological development. For example, a potentially transformative industry may need to be stimulated in the form of government financing for research or for critical infrastructure that lowers barriers to entry. On the other hand, there will be occasions when an emerging technology could spawn an array of societal problems that will need to be addressed before innovative products should be deployed. Compromising safety and responsibility is a ready formulation for inviting crises in which technology is complicit. The harms caused by disasters and the reaction to those harms can stultify technological progress in irrational ways. A central role for ethics, law, and public policy in the development of emerging technologies will be in modulating their rate of development and deployment. While there are financial incentives and regulatory tools for speeding up and slowing down technological development, it would be an illusion to believe this is any easier than controlling an economy. Nevertheless, a GCC can, through advice and influence, play a key role in modulating the development and deployment of new technologies.

Another function of the GCC will be to serve as a central repository for the publicly available science and the research on the social challenges posed by the field of technology it is overseeing. The GCC would not generate new data itself, but would act as a clearinghouse to organize and make available relevant data and studies produced by a variety of other organizations. To the extent funding permits, it might commission its own external reviews or analyses of existing datasets to provide assessments of what is or is not known about a particular technology, or to collect and integrate stakeholder views of concerns, benefits, policies and future projections relating to the technology at issue. The public and the media often have great difficulty in knowing who is credible and whether various potential harms are real or speculative – the GCC may be able to provide snapshots of the technology that can help to discern such issues. While it will not be possible to eliminate the challenges of limited and contested knowledge, it would nevertheless be helpful to have a semi-authoritative committee reporting which harms are imminent and which are highly speculative and dependent on future scientific discoveries. In addition, the GCC would be monitoring when scientific thresholds, which open the doorway to new harms, may be about to be crossed. Members of the press could read reports from the GCC or interview representatives of the GCC as a way of grounding their articles about a wide variety of societal concerns arising from the introduction of new technologies.

In addition to a GCC for each emerging technology, it may be useful to form a governing council for representatives of GCCs for different technologies to compare approaches, successes and failures, and identify cross-technology or convergence issues that might affect multiple technologies. In some cases it may also be necessary for a GCC to coordinate its activities with that of similar bodies in other countries, for example, in initiating and promoting harmonized approaches to governance of specific technologies.

7.5   IMPLEMENTATION ISSUES

While we believe that GCCs would provide a valuable and much-needed mechanism for navigating the challenges posed by emerging technologies, the path to their implementation is far from clear. There are many questions to be answered. From where does the GCC get its power (influence), authority or legitimacy? How are the constituent members and staff of a GCC chosen? How can their credibility be established? To whom is the GCC accountable? Would it be better if a GCC were a government institution or private? From where will its funding come?

While we do not attempt to provide comprehensive answers to these implementation questions here, some preliminary thoughts are provided. The effectiveness of a GCC will be largely determined by perceptions of the GCC's credibility and usefulness to the government, technology developers, non-governmental organizations, the media, and the public. In other words, there is a circularity where influence, authority, and credibility all play off of each other. Certainly all of these could, at least in theory, be built over time. But influence and credibility do not come easily in contemporary societies. It would be helpful if individuals who already have a degree of respect and credibility, such as retired leaders of industry, the military, or academia, took on the establishment of a GCC as a personal mission.

There is no single model for how the GCCs might be funded or managed. One option would be legislation providing funding for creating one or more GCCs, perhaps as a pilot project. Another option would be for a foundation to fund a GCC. Yet a third option would be some sort of combination funding, such as the Health Effects Institute co-funded by the EPA and the auto industry. Finally, funding could be provided by membership fees, primarily paid by industry members as was the case with ICON discussed above. Industry funding alone would probably not be a good model because of the actual or perceived conflict of interest it would have, although even here there may be innovative approaches to structure the industry funding to ensure no undue influence. The funding model for the GCC would influence the selection process for the GCC leadership and the number of staff hired. The selection of a credible and respected leader for a GCC will be critical for establishing its bona fides. Finally, some sort of advisory or steering committee, composed of representatives of relevant stakeholders including government, industry, NGOs, the public, scientists, and governance experts, would also be helpful in building and maintaining credibility and relevance.

The implementation issues can be worked through, though we recognize that these practical issues and concerns might make some believe that establishing an effective and credible GCC in the contemporary political context is either too complicated, hopelessly naïve, or perhaps both. Despite that, modern governments and private institutions have been capable of implementing complicated solutions to difficult problems once the need is fully recognized. Unfortunately, all too often this occurs after a tragedy and the realization that similar calamities will occur if protective measures are not put into place. Such shortsightedness and reactive policy making can easily lead to irrational and overly bureaucratic solutions.

A GCC offers a different and more comprehensive approach to monitor, manage and modulate an emerging technology. As a first step we would recommend implementing a GCC as a pilot project for one or more relatively new technologies that are not yet encumbered by a large network of oversight bodies, existing regulatory programs, and adversarial relations. Examples might include a GCC for robotics or synthetic biology. A vast array of legal and ethical challenges will arise from the introduction of robotic devices for the care of infants, the homebound and the elderly, for military, rescue and surveillance operations, for medical applications, and for entertainment.[30] Altering and synthesizing a new organism poses public health and environmental concerns, and the creation of new forms of life that may offend the ethical sensibilities of various communities. While many of these concerns can be handled within the existing regulatory framework, others pose new problems. For example, Congress asked the FAA to establish test sites for introducing unmanned aerial vehicles into domestic airspace, but the agency stepped back from its deadline of December 31, 2012 stating that privacy issues needed to be resolved before it could revise standards for aerospace safety. Challenges that go beyond the authority of one agency will arise often as innovative technologies are introduced for new contexts and as the emerging technologies converge giving birth to novel possibilities. A new governance model is clearly needed. While there are risks in the GCC approach, we think those risks are minimal and worth taking. A pilot project would offer the opportunity to work out the implementation challenges, as well as the opportunity to witness whether GCCs represent a comprehensive, coherent, and effective means of governing the governance of emerging technologies.

---

[30] Wendell Wallach, "From Robots to Techno Sapiens: Ethics, Law and Public Policy in the Development of Robotics and Neurotechnologies," 3 LAW, INNOVATION AND TECHNOLOGY 185–207 (2011).

PART II

Specific applications

8. The hare and the tortoise: an Australian perspective on regulating new technologies and their products and processes[1]

Diana M. Bowman

"It's a great country, but we've got to be responsible for our actions and it's certainly a bloody nanny state when it comes to what we can do."

Mark Webber, 2010[2]

8.1 INTRODUCTION

The role of government and the limits of state action are often highly controversial. Debates over the so-called "appropriate role" of such intervention, and the limitations of individual rights that such intrusions may give rise to, are woven throughout geography and time. Such controversies have often fallen within the sphere of public health.[3] As highlighted by Jochelson,[4] "[r]egulations we now take completely for granted have sparked fierce controversy in their time"; instruments designed to, for example, govern activities relating to sewage, cigarettes

---

[1] The author is grateful to Professor Jennifer Kuzma, who inspired the title of this chapter.

[2] Paul Millar and David Root, "Webber Blasts Rise of 'Nanny State'," THE AGE, March 29, 2010, available at http://www.theage.com.au/national/webber-blasts-rise-of-nanny-state-20100328-r55n.html (accessed January 28, 2013) (quoting Mark Webber).

[3] See, e.g., F. Adshead and A. Thorpe, "The Role of Government in Public Health: A National Perspective," 121 PUBLIC HEALTH 835 (2007); L. O. Gostin and K. G. Gostin, "A Broader Liberty: J.S. Mill, Paternalism and the Public's Health," 132 PUBLIC HEALTH 214 (2009).

[4] K. Jochelson, "Nanny or Steward? The Role of Government in Public Health," 120 PUBLIC HEALTH 1149, 1150 (2006).

and alcohol consumption. Regulatory instruments designed to alter individual behavior for the purposes of protecting one's own wellbeing, such as mandatory seat-belt and helmet wearing, speed limits, so-called "soda taxes", and the plain packaging of cigarettes continue to be the subject of much debate.[5] As Formula One driver Mark Webber so eloquently reminded us when pulled over by the police for speeding in the state of Victoria in 2010, jurisdictions that adopt a paternalistic approach to public health activities are often negatively labeled as a "nanny state."[6]

This chapter draws upon Webber's critique of Victoria – and Australia in general – as a nanny state when addressing public health problems. It is through this lens that the chapter examines the concept of the "pacing problem" experienced by policy makers and regulators in relation to new technologies and related public health challenges. Australia represents an interesting case study for such a critique, as by all accounts it is a nanny state, and Australia's population enjoys a high standard of living in part due to these public health interventions. The state of Victoria was the first jurisdiction in the world to implement a regulatory framework for a diverse set of technologies and their applications, ranging from mandatory seat belt and motorcycle helmet wearing to access to reproductive technologies such as in vitro fertilization (IVF) and associated embryo research.[7] The encroachment of government in these areas was not without controversy, and signaled for many a step too far in relation to state intervention.

Historically, federal schemes that required prior regulatory approval took a traditional "end-product" approach, rather than focusing on the process by which the product was made. In 2001, a federal regulatory scheme was created to govern certain dealings with genetically modified organisms (GMOs), including the creation of the national Office of the

---

    [5] See, e.g., John G. Adams, RISK AND FREEDOM: THE RECORD OF ROAD SAFETY REGULATION (1985); Robert H. Lustig, Laura A. Schmidt, and Claire D. Brindis, "The Toxic Truth About Sugar," 482 NATURE 27, 28–29 (2012); Tania Voon, Andrew D. Mitchell, Jonathan Liberman, and Glyn Ayres, PUBLIC HEALTH AND PLAIN PACKAGING OF CIGARETTES (2012).

    [6] Millar and Root, supra note 2.

    [7] A. P. Vulcan, M. H. Cameron and W. Watson, "Mandatory Bicycle Helmet Use: Experience in Victoria, Australia," 16 WORLD JOURNAL OF SURGERY 389 (1992); L. Waller, "Australia: The Law and Infertility – The Victorian Experience," in LAW REFORM AND HUMAN REPRODUCTION, 17 (Sheila A. and M. Clean, eds, 1992); C. B. Cohen, "Unmanaged Care: The Need to Regulate New Reproductive Technologies in the United States," 11 BIOETHICS 348 (1997); Jean Cohen et al., "The Early Days of IVF Outside the UK," 11 HUMAN REPRODUCTION UPDATE 439 (2005).

Gene Technology Regulator (OGTR).[8] This introduced a process-focused regulator into the national regulatory landscape. Attention has now been turned to the issue of nanotechnologies and the ability of the federal regulatory matrices to effectively regulate these technologies and their end products. With synthetic biology rapidly becoming a reality, such questions will have to be reframed in the near future to encompass this field and its applications as well.[9] Government intervention in these areas continues to be contentious for some, with particular debate surrounding the appropriate role of government in safeguarding human and environmental health in relation to nanomaterial and nano-objects. For some stakeholders, greater intervention on the basis of public health may be justified at this time. However, for others, such an intrusion when the science remains contested would be going beyond even that which is acceptable in a so-called nanny state.

It is important to note that the question of how to effectively regulate emerging technologies is not new. There is much that can be drawn from earlier experiences, especially from governments and agencies that attempted to pace the technology through formal policy processes and from the consequences of doing so. While some technologies may require the creation of a *sui generis* regulatory framework – or at least be perceived by some to need a novel regulatory approach – the increasingly dynamic nature of innovation suggests that a more generic approach is most appropriate to regulate the forthcoming waves of emerging technologies and their products.

This chapter explores these ideas through an Australian lens, with particular focus on nanotechnologies and the next generation of disruptive technologies. However, it begins by briefly considering two past examples of how Australian governments grappled with the challenges posed by rapid technological developments and what might be learned from them, with specific focus on IVF and GMOs. Both technologies demanded an acute balance between scientific advances at the forefront of knowledge and empathy for broader societal concerns. This chapter then goes on to look at the policy and regulatory landscape for nanotechnologies and the way in which the Australian government has attempted to keep pace with the scientific developments. Unlike GMOs, the government has currently refused to implement *sui generis* legislative

---

[8]   Karinne Ludlow, "Cultivating Chaos: State Responses to Releases of Genetically Modified Organisms," 9 DEAKIN L. REV. 1 (2004).
[9]   See, e.g., E. Stokes and D. M. Bowman, "Regulatory Inheritance and Emerging: Technologies: The Case of Nanotechnologies and Synthetic Biology," 2 EUROPEAN JOURNAL OF RISK REGULATION 235 (2012).

requirements for nanotechnologies. Despite calls for the government to do so, reasons to adopt this slow and steady approach are considered in the final section of this chapter.

8.2 "THERE'S NOBODY IN THE WORLD THAT CAN WIN AGAINST ME," SAID THE HARE

The birth in June 1980 of Candice Reed, Australia's first (the world's third) "test tube baby," signaled a new national era in biomedical science.[10] While this achievement placed a Melbourne-based scientific team at the forefront of new reproductive technologies, the medical advance sparked widespread and fierce public debate around the country.[11] In particular, in conjunction with an important broader religious discourse surrounding the sanctity of human life, this technological advance presented a variety of new moral, ethical and political questions with which governments and policymakers had to grapple.[12]

In response to widespread public debate and concern over IVF research, in 1982 the government of Victoria appointed Professor Louis Waller to chair the Committee to Consider the Social, Ethical, and Legal Issues Arising from In Vitro Fertilization (to become known simply as the Waller Committee). The pioneering work of the Waller Committee sought to better understand the scientific, technological and medical advances of IVF and develop legal, ethical and regulatory frameworks as new medical advances came to light.[13] In short, the committee was charged to develop a *sui generis* regulatory regime that could keep pace with the direction in which the technology was developing. Such a regime was without precedent.

---

[10]   Joseph L. Goldstein, "Laskers for 2001: Knockout Mice and Test-tube Babies," 7 NATURE MEDICINE 1079 (2001); see also Jean Cohen et al., supra note 7.

[11]   Sharyn L. Roach Anleu, "The Legal Regulation of Medical Science," 43 LAW & POLICY 417 (2003).

[12]   LeRoy Walters, "Human Embryonic Stem Cell Research: An Intercultural Perspective," 14 KENNEDY INSTITUTE OF ETHICS JOURNAL 3 (2004).

[13]   The Committee to Consider the Social, Ethical and Legal Issues arising from In Vitro Fertilization, INTERIM REPORT SEPTEMBER 1982 ("Waller Committee"); Victoria Committee to Consider the Social, Ethical and Legal Issues arising from In Vitro Fertilization, REPORT ON DONOR GAMETES IN IVF (1983); The Committee to Consider the Social, Ethical and Legal Issues arising from In Vitro Fertilization, REPORT ON DISPOSITION OF EMBRYOS PRODUCED BY IVF (1984).

Following the recommendations of the Waller Committee, the Victoria Parliament passed the "Infertility (Medical Procedures) Act (Vic)" in 1984, thereby becoming the first jurisdiction in the world to enact legislation to regulate new reproductive technologies such as IVF and associated embryo research.[14] As noted by Roach Anleu, the subsequent proclamation of the "Infertility Treatment Act (Vic)" in 1995 (which repealed the earlier act) provided the state government with a more extensive and detailed framework under which they could regulate reproductive technologies.[15] In this respect, state intervention enabled the government to determine who had the ability to access the technology – including those who were considered medically infertile compared to those who were socially infertile – and under what conditions access would be allowed to approved stakeholders. Victoria's place at the "forefront of [IVF] regulation"[16] is clearly illustrated through the number of other jurisdictions that have subsequently enacted similar (although less extensive) legislation.[17]

Recently, scientific advances in gene technology, including the field of GMOs, have engendered intensive public debate around the world.[18] While countries such as the United States have opted to regulate GMOs under pre-existing legislation,[19] the Australian Government favored a national scheme to regulate potential public health threats such as human and environmental risks associated with the technology.[20] Under the "Gene Technology Act 2000 (Cth)", the Office of the Gene Technology Regulator (OGTR) is provided with the necessary authority to regulate GMOs as a process, effectively establishing what they have called "some

---

[14]   See Waller, supra note 7; C.B. Cohen, supra note 7, Jean Cohen et al., supra note 7.

[15]   See Roach Anleu, supra note 11.

[16]   Infertility Treatment Authority, Annual Report 2 (1999), available at http://www.varta.org.au/secure/downloadfile.asp?fileid=1001242 (accessed November 24, 2012).

[17]   See C. B. Cohen, supra note 7; Jean Cohen et al., supra note 7.

[18]   See BIOTECHNOLOGY – THE MAKING OF A GLOBAL CONTROVERSY (M. W. Bauer and G. Gaskell, eds, 2002).

[19]   Nicholas P. Guehlstorf and Lars K. Hallstrom, "The Role of Culture in Risk Regulation: A Comparative Case Study of Genetically Modified Corn in the United States of America and European Union," 8 ENVIRONMENTAL SCIENCE & POLICY 327 (2005); see also Bauer and Gaskell, supra note 18.

[20]   See Ludlow, supra note 8.

of the toughest regulation[s] in the world concerning biotechnology."[21] However, the OGTR is not the sole regulator of GMOs among federal agencies – others, including the Therapeutic Goods Administration (TGA) and the Food Standards Australia and New Zealand (FSANZ), continued to regulate GM products that fell within their legislative scope prior to the introduction of the national scheme. States similarly have a regulatory role in relation to GMOs. As noted by Ludlow, the matrix of specific state regulatory responses subsequently implemented has "destroy[ed] the uniformity of GMO regulations in Australia."[22] Despite the scientific and societal uncertainties associated with the technology, these actions suggest that it is not without precedent for a federal Australian Government to re-evaluate its approach and overhaul the existing regimes accordingly. This includes specifically regulating a technology by reference to the process itself rather than the end products.

The existence of a *sui generis* regulatory regime for an area of biotechnology has arguably provided a precedent with respect to how other emerging technologies could be governed within Australia. This was particularly the case in the early debates surrounding nano-technologies, which this chapter now considers.

8.3 ANNOYED BY THE BRAGGING, THE AUSTRALIAN GOVERNMENT ACCEPTS THE CHALLENGE

The economic and innovative potential of nanotechnologies to the Australian economy was recognized by the federal government well before the mass commercialization of the technology.[23] Arguably, however, it was not until the publication of the Royal Society and Royal

---

[21] Diana M. Bowman and Graeme A. Hodge, "A Small Matter Of Regulation: An International Review Of Nanotechnology Regulation," 8 COLUMBIA SCI TECH L.J. 1, 23 (2007) (quoting Biotechnology Australia).

[22] Ludlow, supra note 8, at 2.

[23] See Prime Minister's Science Engineering and Innovation Council, NANOTECHNOLOGY – THE TECHNOLOGY OF THE 21ST CENTURY: THE ECONOMIC IMPACT OF EMERGING NANOMETRE SCALE TECHNOLOGY (1999), available at http://www.innovation.gov.au/Science/PMSEIC/Documents/NanotechnologyTheTechnologyOfThe21stCentury.pdf (accessed December 15, 2012); Prime Minister's Science Engineering and Innovation Council, NANOTECHNOLOGY: ENABLING TECHNOLOGIES FOR AUSTRALIAN INNOVATIVE INDUSTRIES (2005), available at http://www.innovation.gov.au/Science/PMSEIC/Documents/Nanotechnology.pdf (accessed December 15, 2012).

Academy of Engineering (RS-RAE) seminal report in the United Kingdom in 2004 that the broader potential human and environmental risks, and the social, ethical and regulatory dimensions of the emerging technology, were considered in depth.

Importantly, at least from the perspective of this chapter, the RS-RAE's mandate included the identification of "areas where additional regulation needs to be considered."[24] As the authors acknowledged, their report was not exhaustive, as the authors focused on the existing regulatory frameworks that they considered as the most challenging at the time in light of the evolving state of the scientific art. These included the regulatory regimes that dealt with, for example, industrial chemicals, occupational health and safety, foods, and therapeutic goods. It is within this focus that the authors noted what they termed a "regulatory gap" in relation to how the industrial chemical regulatory framework would deal with existing chemicals re-engineered at the nanoscale.[25] Concerns were also raised in relation to topics such as the applicability of existing approaches to control exposure within the workplace, the lack of a central repository of nanoscale ingredients being used in consumer products (such as cosmetics), and the need for transparency regarding safety evaluations for consumer products containing nanomaterials. Having identified a number of inherent weaknesses and gaps within the frameworks analyzed, the RS-RAE drew the following conclusion:

> We recommend that all relevant regulatory bodies consider whether existing regulations are appropriate to protect humans and the environment from the hazards outlined in this report, and publish their review and details of how they will address any regulatory gaps.[26]

The UK government acknowledged the need for action, stating in its response to the RS-RAE that, "it is vital that we assess any regulatory gaps at an early stage, to ensure that human health and the environment are adequately protected."[27] However, an explicit commitment of economic resources for such a task was not forthcoming.

---

[24] The Royal Society and Royal Academy of Engineering, NANOSCIENCE AND NANOTECHNOLOGIES: OPPORTUNITIES AND UNCERTAINTIES vii (2004), available at http://www.nanotec.org.uk/report/Nano%20report%202004% 20fin.pdf (accessed December 15, 2012).

[25] Id. at 71.

[26] Id. at xii.

[27] HM Government, RESPONSE TO THE ROYAL SOCIETY AND ROYAL ACADEMY OF ENGINEERING REPORT: 'NANOSCIENCE AND NANO-TECHNOLOGIES: OPPORTUNITIES AND UNCERTAINTIES' 13 (February

The review provided the first in-depth examination of the effectiveness of specific UK and EU regulatory regimes when dealing with nano-materials and nano-based consumer products, and the potential limits thereof. This report was timely, as the findings added to the growing scientific literature, enabled commentators to move away from the simple rhetoric underpinning many of the debates, and provided stakeholders with the necessary details to more carefully articulate the most pressing areas of concern. In short, the spotlight had been firmly turned on to the emerging regulatory challenges that nanotechnologies would present to governments and safety regulators. And while the RS-RAE's report focused on the EU context, the ripples were felt as far away as Australia.

Given the then-Howard government's decision to create a national scheme for GMOs, and the moratoriums that had been established by multiple states in relation to the commercial release of GMOs,[28] it is hardly surprising that a number of stakeholders argued for the establishment of an analogous scheme for nanotechnologies.[29] Alongside this were calls for a moratorium on the commercialization of the technology until adequate data was available for validated risk assessments to take place.[30] The precedent, it would seem for such stakeholders, had been established.

Such calls appear to have fallen on deaf ears. Once the hare, the Australian government appeared more comfortable taking a tortoise-like approach to policy and regulatory decisions relating to this latest emergent technology. That is not to say that the landscape remained unchanged, as various government institutions have sought to advance policy within the industry. Such examples include the Department of Industry, Tourism and Resources (DITR), which established the National

---

2005), available at http://webarchive.nationalarchives.gov.uk/+/http://www.dti. gov.uk/files/file14873.pdf (accessed December 1, 2012).

[28]   See Ludlow, supra note 8.

[29]   See International Center for Technology Assessment, COMMENTS ON PRIORITIES FOR THE TRANSATLANTIC INNOVATION DIALOGUE (2010), available at http://www.state.gov/documents/organization/137963.pdf (accessed November 18, 2012)

[30]   See Friends of the Earth Australia, NANOMATERIALS, SUNSCREENS AND COSMETICS: SMALL INGREDIENTS, BIG RISKS (2006), available at http://libcloud.s3.amazonaws.com/93/ce/0/633/Nanomaterials_sunscreens_and_ cosmetics.pdf (accessed March 3, 2013); Standing Committee on State Development, NANOTECHNOLOGY IN NEW SOUTH WALES (2008), available at http://www.parliament.nsw.gov.au/Prod/parlment/committee.nsf/0/35d2e3e37498 a908ca2574f1000301bb/$FILE/Final%20Report%20Oct.pdf (accessed November 11, 2012)

Nanotechnology Strategy Taskforce (NNST) in July 2005 to devise "options for a coordinated, national strategy on nanotechnology,"[31] or the Senate Community Affairs Committee Inquiry, which included language to investigate the potential effects of emerging technologies, including nanoparticles, with respect to workplace exposure of toxic dust.[32] An independent review of federal regulatory arrangements was also commissioned by the NNST in 2006. The adoption of a "whole-of-government" approach to nanotechnologies included the establishment of a cross-agency Health, Safety and Environment (HSE) Working Group,[33] influenced the scope of the review. Pursuant to the Terms of Reference (ToR), the consultants were required to analyze the adequacy of *all relevant federal health and safety regulatory frameworks* for nanotechnologies, as well as to examine the existing intellectual property (IP) regimes.[34] While the consultants were requested to identify gaps in these frameworks, they were expressly informed that they were not to "make recommendations on addressing these gaps."[35]

The report was handed to the HSE Working Group in June 2007. As the authors of the report acknowledged, it was "the result of a regulatory terrain mapping and regulatory analysis exercise rather than a definitive study of the application of particular parts of the relevant regulatory frameworks or a comprehensive analysis of all federal regulatory regimes that could be impacted by nanotechnology."[36] It is not the intent of this chapter to critique the key findings of the Australian review, or provide

---

[31] National Nanotechnology Strategy Taskforce, OPTIONS FOR A NATIONAL NANOTECHNOLOGY STRATEGY – REPORT TO MINISTER INDUSTRY, TOURISM AND RESOURCES (2006), available at http://catalogue.nla.gov.au/Record/4314929 (accessed December 1, 2012).

[32] The Senate, COMMUNITY AFFAIRS REFERENCES COMMITTEE – WORKPLACE EXPOSURE TO TOXIC DUST – TERMS OF REFERENCE 15 (2006), available at http://www.aph.gov.au/Parliamentary_Business/Committees/Senate_Committees?url=clac_ctte/completed_inquiries/2004-07/toxic_dust/report/report.pdf (accessed December 1, 2012).

[33] Comprised of members from federal regulatory bodies, federal bureaucrats and members of the scientific community.

[34] Karinne Ludlow, Diana M. Bowman, and Graeme A. Hodge, A REVIEW OF THE POSSIBLE IMPACTS OF NANOTECHNOLOGY ON AUSTRALIA'S REGULATORY FRAMEWORKS (2007), available at http://www.innovation.gov.au/Industry/Nanotechnology/NationalEnablingTechnologiesStrategy/Documents/MonashReport2007.pdf (accessed March 14, 2013).

[35] Id. at 9.

[36] Id.

any in-depth commentary on the terrain covered by its authors. For the purposes of this chapter, it is instead important to note that while the Australian government did not formally respond to the review,[37] the findings appear to have had an impact on Australian regulatory and policy discourse. In assessing the impact of the review, Bowman and Ludlow found that:

> all of the regulatory schemes response for end products – agricultural chemicals, veterinary medicines, food / food packaging, therapeutic goods – had responded to the gaps identified by Ludlow et al. in one manner or another. The regulator responsible for workplace safety ha[d] also responded.[38]

These authors note that such responses have occurred within an evolving political and regulatory environment in which some regulators have seen extensive reviews and changes to their respective schemes, which occurred independently of the review.[39]

In order to reduce the regulatory burden and make regulation "better", such reform has been a hallmark of the broader policy landscape within Australia and other developed economies over the last few decades.[40] The so-called "hollowing out" of the state, as so eloquently coined by Rhodes, appears to be *prime facie* in conflict with the "nanny-state"

---

[37]   A formal response was however provided by the Australian Office of Nanotechnology (2010), in their 2008–09 Annual Report (See Australian Office of Nanotechnology, 2008–09 ANNUAL REPORT (2010), available at http://www.innovation.gov.au/Industry/Nanotechnology/NationalEnablingTechnologies Strategy/Documents/National_Nanotechnology_Strategy_AR2008-09.pdf (accessed December 1, 2012).

[38]   Diana M. Bowman and Karinne Ludlow, "Assessing the Impact of a 'For Government' Review on the Nanotechnology Regulatory Landscape," 38 MONASH LAW REVIEW 3, 168–212 (2013).

[39]   Id.

[40]   Brian Head, DEREGULATION OR BETTER REGULATION? ISSUES FOR THE PUBLIC SECTOR (1991); OECD, FROM RED TAPE TO SMART TAPE: ADMINISTRATIVE SIMPLIFICATION IN OECD COUNTRIES (2003), available at http://www.oecd.org/regreform/regulatory-policy/2790042.pdf (accessed December 1, 2012); HM Government, REDUCING REGULATION MADE SIMPLE: LESS REGULATION, BETTER REGULATION AND REGU-LATION AS A LAST RESORT (2010), available at https://www.gov.uk/government/uploads/system/uploads/attachment_data/file/31626/10-1155-reducing-regulation-made-simple.pdf (accessed December 1, 2012).

label.[41] However, regulatory reform would appear to be as much about improving procedures, efficiencies, access, and transparency, and reducing costs rather than reducing regulatory activity.[42] While the move towards deregulation may be associated with a reduction in command and control regulation, it is also appears to be correlated with an increase in non-state actors within the regulatory paradigm.[43] As such, the void left by the state's withdrawal appears to be filled by new governance arrangements.[44] Arguably less hierarchical in nature, and characterized by greater fragmentation and decreased formal legal authority, these new governance arrangements nevertheless deploy a vast number of regulatory tools through which they shape and alter behavior.[45] More flexible in nature, such new governance arrangements appear to be preferable within complex and rapidly evolving areas such as emerging technologies. This is in part due to the multiplicity of actors and instruments that emerging technologies deploy, and to the ability to harness the creativity and expertise of both public and private actors.

Against the background of this broader regulatory trend, arguably the most significant whole-of-government impact regarding the federal nano-technologies policy landscape was the implementation of the National Enabling Technologies Strategy (NETS).[46] Housed within the Department of Innovation, Industry, Science and Research (DIISR), the four

---

[41] R.A.W. Rhodes, "The Hollowing out of the State: The Changing Nature of the Public Service in Britain," 65 THE POLITICAL QUARTERLY 138 (1994).

[42] Steven Kent Vogel, FREER MARKETS, MORE RULES: REGULATORY REFORM IN ADVANCED INDUSTRIAL COUNTRIES (1996).

[43] Bridget M. Hutter, THE ROLE OF NON-STATE ACTORS IN REGULATION, CARR DISCUSSION PAPER SERIES NO. 37 (April 2006), available at http://eprints.lse.ac.uk/36118/1/Disspaper37.pdf (accessed December 4, 2012); Neil Gunningham, "The New Collaborative Environmental Governance: The Localization of Regulation," 36 J.L. SOC'Y 145 (2009).

[44] THE POLITICS OF REGULATION: INSTITUTIONS AND REGULATORY REFORM FOR THE AGE OF GOVERNANCE (Jacint Jordana and David Levi-Faur, eds., 2004).

[45] Neil Gunningham, "Environmental Law, Regulation and Governance: Shifting Architectures," 21 JOURNAL OF ENVIRONMENTAL LAW 179 (2009); Arie Freiberg, THE TOOLS OF REGULATION (2010).

[46] Australian Government, NATIONAL ENABLING TECHNOLOGIES STRATEGY (2009), available at http://www.innovation.gov.au/Industry/Nanotechnology/NationalEnablingTechnologiesStrategy/Documents/NETS_booklet.pdf (accessed December 2, 2012).

year, \$A38.2 million strategy supersedes the National Nanotechnology Strategy.[47] The new strategy:

> provides a framework to support the responsible development of enabling technologies. Its aim is to improve the management and regulation of biotechnology and nanotechnology in order to maximise community confidence and community benefits from the use of new technology.[48]

Rather than focusing on a specific technology, the focus on "enabling technologies" is significant. For starters, this focus provides the strategy with the ability to better pace the emergence of new technologies as such advances are likely to fall, by default, within its remit. Second, the implementation of an overarching Strategy negates the need for continual reinvention. Focus can instead be placed on development of new policies that keep pace with the technology, rather than on the establishment of a new office, agency, or strategy itself. Third, with the increasing convergence of technologies – bio-nanotechnology, neural and cognitive technologies and synthetic biology, to name just a few – the complexities of developing new policies and appropriate regulatory instruments will only increase. Traditional scientific and disciplinary boundaries are increasingly blurred and there is an inherent need for diverse expertise to be located in one spot in order to ensure that any such developments reflect the evolving states of the scientific art and community concerns.

NETS is underpinned by six themes and objectives. Two of these – *balancing risk and reward* and *planning for the future* – incorporate or support policy and regulatory objectives. Pursuant to the *planning for the future* theme, "the strategy will assist government ... to prepare for the advent of new technologies by undertaking foresighting activities and supporting the development of policy and regulatory frameworks."[49]

To assist in fulfilling these objectives, two external boards were established under the strategy: a Stakeholder Advisory Council, whose mandate is to draw the government's attention to specific issues or areas of concern regarding enabling technologies in the community, and the Expert Forum, whose members are charged with "identify[ing] new and converging technologies to inform government policy and strategy, and to improve how science and technology are utilized by industry, government

---

47   Id.
48   Id. at 3.
49   Id. at 4.

and society in general."[50] This model is arguably reflective of the broader "better regulation" movement. External expertise has been explicitly incorporated into the strategy, thereby promoting greater collaboration amongst the different sectors, while also promoting transparency and legitimacy. Through the two expert bodies, the state has the ability to harness the strengths and knowledge of key stakeholders including, for example, the perceptions and concerns of the public, and the barriers, challenges or regulatory concerns being faced by industry. Direct access to industry, civil society, and academia arguably strengthens the capacity of the state to respond to emerging governance issues and concerns.

NETS is further supported by the HSE Working Group, comprised of members from federal regulatory bodies, federal bureaucracy, and the scientific community. The HSE Working Group assists the government with the coordinated approach to emerging technologies, but also provides its members with a reference point for national and international developments. This working group would appear to be a strength of the strategy, as it draws together those with the scientific, policy, and regulatory expertise into one forum in order to consider the complexities, uncertainties, and trajectories of the technologies. It provides the department with the expertise to better pace the scientific advances. One can hypothesize that this type of forum works well in a country such as Australia, but arguably less so in larger countries. Due to the lower churn rate and relatively small size of the regulatory bodies and the scientific community, members of the working group generally have a longstanding record of working with each other. The concentrated nature of these bodies and individuals within three cities would also appear to promote informal opportunities for information exchange. While it is not possible to measure the impact that each of these three bodies have had to date on the policy landscape, their mere existence suggests that the Australian government is aware of the need for high level expert input on emerging technologies in terms of the science, policy, and legitimacy of its actions.

Since its inception, the NETS Office has coordinated numerous foresighting, communication, engagement, and educational activities.[51] The majority of these have been uncontroversial in nature. However, the

---

[50] Australian Government, TERMS OF REFERENCE FOR THE NETS EXPERT FORUM (2010), available at http://www.innovation.gov.au/Industry/Nanotechnology/NationalEnablingTechnologiesStrategy/Documents/TermsofReferenceExpertForum.pdf (accessed December 4, 2012). The author of this chapter was a member of the Expert Forum (2010–2012).

[51] National Enabling Technologies Strategy, 2009–10 ANNUAL REPORT (2010), available at http://www.innovation.gov.au/Industry/Nanotechnology/

commissioning of an independent study to examine the feasibility of an Australian mandatory reporting registry has sparked significant controversy and commentary from key Australian stakeholders.[52] The scope of the study was clear: to "examine the feasibility of a mandatory nanotechnology product registry and issues related to the possible implementation of such a registry."[53] Underpinning this objective was the question of whether or not such a registry would assist in addressing any so-called gaps within the currently federal regulatory matrix.[54] While noting the limitations of undertaking such as study – for example, the absence of "hard evidence of technical feasibility"[55] – the study concluded that such a registry was unwarranted in light of the current state of the scientific art, and that the existing federal regulatory matrix was sufficiently robust to address perceived risks associated with the technology.[56] In their words, the introduction of a mandatory registry "could involve substantial costs but provide few or questionable benefits ... ."[57] With key activists such as Friends of the Earth (FoE) Australia continuing to call upon the Australian government for compulsory labeling,[58] this issue appears

---

NationalEnablingTechnologiesStrategy/ Documents/NETSAnnualReport200910. pdf (accessed December 4, 2012).

[52]   See Friends of the Earth Australia, EXPOSURE TO NANOMATERIALS CAN HAVE A SERIOUS HEALTH IMPACT (2012), available at http:// nano.foe.org.au/exposure-nanoparticles-can-have-serious-health-impacts (accessed December 1, 2012); Rachel Carbonell, "Government Resists Calls for a Nanomaterials Registry," ABC NEWS, July 25, 2012, available at http://www.abc. net.au/news/2012-07-25/calls-for-improved-nano-technologies-regulation/41537 78 (accessed December 1, 2012).

[53]   Centre for International Economics, FEASIBILITY OF IMPLEMENTING A MANDATORY NANOTECHNOLOGY PRODUCT REGISTRY 7 (2011), available at http://www.innovation.gov.au/Industry/Nanotechnology/National EnablingTechnologiesStrategy/ Documents/ FeasibilityMandatoryNanotechProduct Registry.pdf (accessed January 2, 2012).

[54]   Id.

[55]   Id. at 8.

[56]   Id. at 63.

[57]   Id. at 7.

[58]   See, e.g., Rachel Carbonell, "More Consumer Support for Compulsory Labeling of Sunscreens with Nano-particles," ABC News, December 1, 2011, available at http://www.abc.net.au/news/2011-12-01/more-consumer-support-for-compulsory-labelling-of/3707142 (accessed January 2, 2012); Friends of the Earth Australia, NEW ZEALAND TO LABEL NANO-INGREDIENTS IN COSMETICS – SO WHY CAN'T AUSTRALIA?, (2012), available at http:// nano. foe.org.au/new-zealand-introduce-nano-ingredients-cosmetics-–-so-why-can't-australia (accessed January 2, 2012).

unlikely to disappear from the policy debates in the short to medium term. It is worth noting that despite the often high-profile activism, Miller and Scrinis have claimed that the NGO community has had "limited political leverage in nanotechnology debates."[59]

These policy activities have been further supplemented by activities within the federal regulatory agencies themselves. Such policy activities are discussed in greater length below.

8.4 SLOW AND STEADY: THE TORTOISE'S APPROACH TO REGULATING UNKNOWN RISKS

To date, the Australian government has taken a far less aggressive regulatory response to nanotechnologies than previous emerging technologies, such as biotechnologies. However, that is not to say that the *status quo* has continued over the last five years in relation to how the federal regulatory agencies perceive and address nanotechnology-based products and processes. But the approach adopted so far is clearly more analogous to that of the tortoise than the hare.

Arguably, the key finding of the Australian review is the one statement that did not appear anywhere in its text: there is no need to panic. That is not to say that the regulatory arrangements were deemed to be perfect, and that no further action was required. Rather, the report identified six regulatory triggers that *may fail to fire* when having to deal with nanotechnologies. As noted by Ludlow et al.,

> Whilst there is no immediate need for major changes to the regulatory regimes, there are many areas of our regulatory regimes which, potentially, will need amending, and this will be a long term effort across multiple regulators and regulatory agencies as nanoproducts arise and as new knowledge on hazards, exposure and monitoring tools becomes available.[60]

As noted in Section 8.3, Australia's key federal safety regulators have responded to the findings of the review in one capacity or another. Detailing the extent of the policy and regulatory activities on an

---

[59] G. Miller and G. Scrinis, "The Rise of NGOs in Governing Nanotechnology: Beyond the Benefits Versus Risks Framing," in INTERNATIONAL HANDBOOK ON REGULATING NANOTECHNOLOGIES 409, 438 (G. Hodge, D. M. Bowman and A. D. Maynard, eds, 2010).

[60] Ludlow et al., supra note 34, at 4.

agency-by-agency basis is beyond the scope of this chapter.[61] Rather, as highlighted by Bowman and Ludlow, there has been a diversity of responses among the agencies, with some regulators such as those responsible for pesticides and veterinary medicines, therapeutic goods, workplace safety, foods and food packaging and industrial chemicals having directly responded to the gaps identified by the review in one manner or another.[62] In addition to a number of agencies undertaking their own internal reviews and developing strategy documents, specific examples include:

- two voluntary data call-ins for industrial nano-materials (by NICNAS);[63]
- development of a Regulatory Reform of Industrial Nanomaterials strategy, which has included the adoption of new administrative procedures for the notification and assessment of new industrial nanomaterials, as defined by the adoption of a working definition of an "industrial nanomaterial" (by NICNAS);[64]
- new administrative procedures for certain regulated foods that contain nanomaterials requiring prior assessment and approval (by FSANZ);[65]
- a review of the scientific literature on the safety of nanomaterials in sunscreens (by the TGA);[66] and

---

[61]   See, e.g., id. (detailing the nano-specific action by each of the key Australian regulatory bodies in-depth since the 2007 Review).

[62]   Id.

[63]   NICNAS, SUMMARY OF 2008 CALL FOR INFORMATION ON THE USE OF NANOMATERIALS (2010), available at http://www.nicnas.gov.au/ Publications/Information_Sheets/General_Information_Sheets/NIS_Results_ Call_for_Information_2008_Nov_2010_PDF.pdf (accessed December 4, 2012).

[64]   NICNAS, GUIDANCE ON NEW CHEMICAL REQUIREMENTS FOR NOTIFICATION OF INDUSTRIAL NANOMATERIALS (2010), available at http://www.nicnas.gov.au/current_issues/Nanotechnology/Guidance%20on%20 New%20Chemical%20Requirements%20for%20Notification%20of%20Industrial %20Nanomaterials.pdf (accessed December 4, 2012).

[65]   Nick Fletcher and Andrew Bartholomaeus, "Regulation of Nanotechnologies in Food in Australia and New Zealand," 1 INTERNATIONAL FOOD RISK ANALYSIS JOURNAL 33 (2011).

[66]   Department of Health and Ageing Therapeutic Goods Administration, A REVIEW OF THE SCIENTIFIC LITERATURE ON THE SAFETY OF NANO-PARTICULATE TITANIUM DIOXIDE OR ZINC OXIDE IN SUNSCREENS (2009), available at http://www.safersunscreens.com/downloads/sunscreen-zotd. pdf (accessed December 4, 2012).

• the publication of a number of guidance documents to assist with better management of potential risks associated with nanomaterials in the work place (by SWA).[67]

In contrast, some regulatory agencies focused primarily on regulating activities rather than end-products have, understandably, taken a less active approach. Regulators including the OGTR and those responsible for the environment and imports/exports were found to be less active in responding to the findings of the review over the past five years.[68] Such agencies were also found to have been less likely to be active and evident players in the policy debates. However, this is not surprising given that these schemes were found to have fewer gaps than those responsible for the regulation of end-products such as cosmetics and foods.

In an effort to regulate nanotechnologies, the Australian government and its agencies have differed from the European Parliament in that Australia has resisted wholesale changes to regulatory regimes. Until now, it has been content to tweak existing administrative procedures within a few agencies (FSANZ and NICNAS), and to develop guidance material for areas such as worker safety. The adoption of nano-specific provisions in the Cosmetic Regulation and the Food Information to Consumers Regulation by the European Parliament and Council,[69] in concert with the European Commission's adoption of a definition for "nanomaterials"[70] has, without question, signaled that the EU intends to be the hare in this regulatory race. Along with other jurisdictions such as the US, Australia is seemingly analogous to the tortoise, plodding along the path. This is despite vocal and high profile activism to speed up the implementation of nano-specific governance arrangements by certain stakeholder groups.

---

[67]   Bowman and Ludlow, supra note 38.

[68]   Id.

[69]   Diana M. Bowman, Geert van Calster and Steffi Friedrichs, "Nanomaterials and the Regulation of Cosmetics," 5 NATURE NANOTECHNOLOGY 92 (2010), Jean-Philippe Montfort, Sebastien Louvion and Leticia Lizardo, NEW EU FOOD LABELING REGULATION PUBLISHED (2011), available at http://www.mayerbrown.com/files/Publication/a8878b90-17fa-4495-a3b1-6aeafafdc21 2/Presentation/PublicationAttachment/28487465-8029-4295-81ee-6c44471d530 0/11870.pdf (accessed December 4, 2012).

[70]   The European Commission, "Commission Recommendation of 18 October 2011 on the Definition of Nanomaterial," 275 OFFICIAL JOURNAL OF THE EUROPEAN UNION 38, 38–9, available at http://eur-lex.europa.eu/Lex UriServ/ LexUriServ.do?uri=OJ:L:2011:275:0038:0040:EN:PDF (accessed December 14, 2012).

8.5   WHO WILL WIN THE REGULATORY RACE?

There are many different hypotheses that could be constructed as to why the Australian government has adopted its current approach to regulate nanotechnologies. Arguably, none of these will ever fully explain the decision making that has been undertaken behind closed doors. Nor is the government likely to ever adequately control the multitude of confounders associated with such decision making. Placing these caveats aside, there are a number of observations that can be made in relation to why the Australian government has taken a less aggressive approach to the regulatory challenges posed by nanotechnologies than one may have predicted given the precedent set in relation to GMOs.

The fact that the strategy has been developed and implemented by the Department of Innovation, Industry, Science and Research and Tertiary Education appears to be an important factor here. The focus of the department is – understandably – on:

> shap[ing] the future economy, through skills, learning, discovery and innovation. [The] department, and the wider portfolio, are working to accelerate productivity growth and secure Australia's prosperity in a competitive global economy.[71]

Strategic areas within the department include, for example, ensuring the future of Australia's manufacturing and automobile industry, promoting clean technologies, and driving innovation within the research community.[72] The focus on securing the economic benefits of nanotechnologies for Australian industry and research domains are further highlighted by the NETS's commissioning of the Australian Academy of Science (AAS) to produce a *National Nanotechnology Research Strategy*.[73] As noted in the text of the report itself, "there are exciting research

---

[71]   Department of Industry, Innovation, Science and Research and Tertiary Education, ABOUT US – OUR DEPARTMENT (2012), available at http://www.innovation.gov.au/ABOUTUS/OURDEPARTMENT/Pages/default.aspx (accessed December 14, 2012).

[72]   Id.

[73]   Australian Academy of Science, NATIONAL NANOTECHNOLOGY RESEARCH STRATEGY (2012), available at http://www.science.org.au/policy/documents/nanotech-research-strategy.pdf (accessed December 14, 2012). The author of this chapter was one of the more than 50 contributors to the National Nanotechnology Research Strategy report.

and commercial opportunities that grow out of Australian nano-technology strengths across much of the current research leadership."[74] With this in mind, it appears unlikely that the Department of Innovation would wish to hamper commercial opportunities associated with the technology when the state of the science is evolving and potential risks remain highly debated and contested.

Importantly, none of the key federal regulatory bodies concerned with health and safety dimensions of nanotechnologies falls within the Depart-ment of Innovation's umbrella. However, that is not to say that the minister, department secretary or members of the strategy are not concerned with the health and safety dimensions of the technology, as the very fact that such aspects fall within the ambit of the strategy and the HSE Working Group is testament to this. Rather, it suggests that there is a potential disconnect between those generating policy through the NETS program and within the Department of Innovation, and those responsible for driving policy and regulatory reform in those departments where key safety regulators sit. This includes, for example, the Department of Health and Ageing, under which the TGA, OGTR, NICNAS and FSANZ sit.

The evolving state of the science and the acknowledged knowledge gaps in relation to potential risks of some nanomaterials under specific and narrow conditions have been well documented.[75] So too are the concerns relating to the appropriateness of conventional risk assessment paradigms for certain free nanomaterials, such as non-biodegradable/insoluble nanoparticles like metal oxides.[76] A number of significant collaborative research projects have been established, many of which bridge jurisdictional and sectorial boundaries in order to address these concerns and generate the data that is so obviously needed.[77] While

---

[74]   Id. at 4.

[75]   See, e.g., Andrew D. Maynard et al., "Safe Handling of Nanotechnology," 444 NATURE 267 (2006); R. D. Handy, R. Owen and E. Valsami-Jones, "The Ecotoxicology of Nanoparticles and Nanomaterials: Current Status, Knowledge Gaps, Challenges and Future Needs," 17 ECOTOXICOLOGY 315 (2008).

[76]   See, e.g., Milind Kandlikar et al., "Health Risk Assessment for Nano-particles: A Case for Using Expert Judgment," 9 NANOTECHNOLOGY AND OCCUPATIONAL HEALTH 137 (2007); Scientific Committee on Consumer Products, OPINION ON SAFETY OF NANOMATERIALS IN COSMETIC PRODUCTS (2007), available at http://ec.europa.eu/health/ph_risk/committees/04_sccp/docs/sccp_o_123.pdf (accessed December 7, 2012).

[77]   See, e.g., R. J. Aitken et al., EMERGNANO: A REVIEW OF COM-PLETED AND NEAR COMPLETED ENVIRONMENT, HEALTH AND SAFETY RESEARCH ON NANOMATERIALS AND NANOTECHNOLOGY

legislators within the EU have viewed these scientific gaps as one of the key reasons to move forward with legislative measures for regulating nanotechnologies, federal safety regulators within the Australian context appear to prefer to keep a watching brief on the science as it unfolds. One can speculate that greater certainty in relation to public health risks will be needed before the Australian policy and regulators will be willing to know which of these approaches was the superior option. However, it is important to recognize push for legislative change to the various schemes.

Only in hindsight will we discover that there are actually two races currently being run in relation to nanotechnologies. The first race is the pacing race, in which the technology is pitted against those who seek to effectively regulate it. The second race is the regulatory race, in which national and supranational governments are competing among themselves in the development and implementation of policy, administrative and regulatory responses to the technology. In relation to this second race, greater divergence is being observed as these responses develop.[78]

Scientific uncertainty alone would not appear to be a barrier to regulatory reform in the Australia context; the implementation of the OGTR scheme is testament to this. Given the extent of the unknowns and the breadth of the potential applications and end products that nanotechnologies will contribute to, pushing ahead with legislative reform at this time may lock safety regulators into what is ultimately an undesirable position. Arguably, a more strategic approach at this time is to focus on amending administrative procedures and the creation of softer guidance materials, so as to ensure best practice as reflected at the time. Such an approach is inherently more flexible than reliance on legislative instruments, and can be amended in a more judicious manner so as to better

---

(2009), available at http://randd.defra.gov.uk/Document.aspx?Document=CB0409_7910_FRP.pdf (accessed December 7, 2012); INTERNATIONAL HANDBOOK, supra note 59; Gary E. Marchant, Kenneth W. Abbott, Lyn M. Gaudet and Douglas J. Sylvester, "Transnational New Governance and the International Coordination of Nanotechnology Oversight," in THE NANO-TECHNOLOGY CHALLENGE: CREATING LEGAL INSTITUTIONS FOR UNCERTAIN CHALLENGES (David Dana, ed., 2011).

[78] See Linda Breggin, Robert Falkner, Nico Jaspers, John Pendergrass & Read Porter, SECURING THE PROMISE OF NANOTECHNOLOGIES: TOWARDS TRANSATLANTIC REGULATORY COOPERATION (2009), available at http://www.chathamhouse.org/sites/default/files/public/Research/Energy,%20Environment%20and%20Development/r0909_nanotechnologies.pdf (accessed December 7, 2012); Maynard, supra note 75.

keep pace with the evolving state of the science and the commercial reality in which the government finds itself.

Indeed, it would appear that the approach being adopted at this time by the Australian government and its regulatory agencies is one focused on scientific principles rather than tightly construed definitions, many of which remain controversial.[79] By focusing on novelty, likelihood of risk, plausibility and impact of harm to human and/or environmental health, policy and regulatory actors may be better positioned to develop and implement a tailored triage system that addresses the very novelty associated with the technology. Such a principle-based approach may circumvent the adoption of contested definitions, and due to the technology-neutrality of such an approach, may be more readily deployed for other emerging technologies as they make their way into the market place. Such an approach would appear to reflect the federal government's approach to looking at emerging technologies generally, rather than focusing on specific technologies *per se.*

History shows that Australian governments have often taken a paternalistic approach towards public health issues, including the regulatory regime of technologies and their end products. The Australian population enjoys a high quality of life at least in part due to the nanny-state culture that underpins many everyday activities. Such an approach is not without its detractors or controversy. For some, it may therefore seem surprising that the Australian government has taken a more reserved approach to regulating nanotechnologies. Rather than implement sweeping regulatory reforms, policy and safety regulators have attempted to pace scientific and commercial developments through active participation in projects driven by transnational organizations, in-house reviews, and development of administrative and guidance documents that reflect the evolving state of the scientific art. Given the moving target that they must regulate and the competing interests at play within the constituencies, striking a balance between under- and over-regulation will continue to challenge these bodies. Only in time will we be able to determine with any degree of certainty whether this tortoise-like approach to the race was the right tactic to adopt.

---

[79] Andrew D. Maynard, "Don't Define Nanomaterials," 475 NATURE 31 (2011).

9. Properly paced? Examining the past and present governance of GMOs in the United States

Jennifer Kuzma

9.1 INTRODUCTION

A case study of genetically modified organisms (GMOs)[1] in US agriculture and the environment illustrates the problem of policy systems to keep up or pace with advances in emerging technologies. This chapter describes the history of GMO governance in four phases, examining the oversight system's ability to pace with technological developments in each phase. In general, government decisions for oversight of GMOs, particularly GM crops, seemed to pace well with technology in a temporal sense. However, they continue to be contested and do not seem appropriate in the longer term for ensuring safety, transparency and public confidence. The GM crop oversight system exhibited temporal pacing through flexible legal frameworks, but not proper pacing. This chapter argues for a broader notion of pacing that incorporates not only elements of timeliness, but also notions of appropriateness in dynamic societal contexts. It will conclude with proposed lessons from the US GMO oversight experience for developing a new prototype model of governance for emerging technologies that properly paces with technological advancements. This model is based upon three pillars: (i) upstream oversight assessment (a subset of anticipatory governance); (ii) dynamic oversight; and (iii) strong objectivity through more extensive public and stakeholder engagement in decision making.

---

[1] Natural scientists prefer the term genetically engineered; however, we use genetically modified (GM), as it is more in line with international policy discussions. We use GM to indicate any organism modified by recombinant DNA or newer biotechnology methods.

9.2 THE PACING PROBLEM CONSTRUCTED

Scholars, including many of this book's authors, have pointed out the inability of legal and regulatory frameworks to keep pace with technologies.[2] These situations can cause over- or under-reach in ensuring the health and safety of new technological products, which can stall technological development or allow it to proceed unchecked with negative societal consequences. Gaps in product review or the placement of new products in old, byzantine regulatory systems create mistrust in and discontent with systems of governance. This general phenomenon has been described as the pacing problem. Governance for emerging technologies lags behind technological innovation, and innovation in governance needs to match technological innovation.

To further understand the issue of pacing and why it is a problem, historical oversight case studies can be instructive.[3] The story of oversight for GMOs in agriculture provides such a case study. The development of GMO technologies in the United States has proceeded over a 40-year period (see Figure 9.1).[4] The oversight of this developing technology has proceeded in four phases, each with different features of pacing in a temporal sense (see Figure 9.2). We have previously identified three of these phases[5] and review them here in the context of the pacing problem: evolution, implementation and adaptation. For the first time, we describe the most recent oversight phase, revolution, and discuss it in the context of pacing. Each oversight phase is distinct in its ability to keep up with GMO technologies and products and to adjust policies and procedures. Notably some of the most effective phases in rapidly changing oversight matching new GMOs present the greatest problems with regard to market and public failures. Thus, the case study suggests a need to redefine pacing to not only keep up with technology, but also to stay current with evolving public concerns, hopes and discontents about the technology. Suggestions are made for three innovations in governance to ensure not only pacing, but societally responsive and responsible pacing, which we term "proper pacing."

---

[2] THE GROWING GAP BETWEEN EMERGING TECHNOLOGIES AND LEGAL-ETHICAL OVERSIGHT: THE PACING PROBLEM (Gary E. Marchant, Braden R. Allenby and Joseph R. Herkert, eds, 2011).

[3] Jennifer Kuzma, Joel Larson and Pouya Najmaie, "Evaluating Oversight Systems for Emerging Technologies: A Case Study of Genetically Engineered Organisms," 37 J.L. MED. & ETHICS 546 (2009).

[4] Id.

[5] Id.

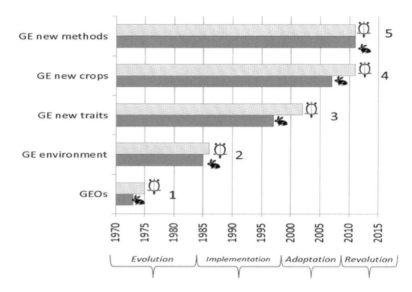

Note: Technological development and deployment are presented by the solid bars (rabbit icon), and key federal oversight advancements are represented by the shaded bars (turtle icon). GEOs = production of GEOs in laboratory; GE environment = release of GEOs in environment; GE new traits = release of crops with pharmaceutical or industrial traits; GE new crops = alfalfa, sugar beet, amylase corn, bluegrass; GE new methods = targeted genetic modification, etc. For example, at point 3, pharmaceuticals in crops were released in the environment in 1997, while specific guidance on isolation distances in 2003.

Figure 9.1   Key Innovation and Related Oversight Activity Over Time

The tortoise and the hare fable illustrates the pacing of the governance system (the tortoise) with various GM technological advances (the hare) in the four phases. In this fable, the hare becomes over-confident and stops to rest, while the tortoise maintains a steady pace and ultimately wins the race. However, the use of the analogy does not imply a winner of the race, but rather the stopping and starting of the hare and the slow pace of the tortoise. The tortoise stands for oversight. Traditionally the United States has governed GM crops through changes in regulations, policies and procedures, but with significant delays in their revision and execution. Comment periods and legal challenges can significantly slow government. The hare symbolizes technological development and deployment, as these will move more quickly unless the industry stops itself or is delayed by government. In other words, the normal operating mode of government oversight is slow, while the normal operating mode of the biotechnology industry is fast, even though external forces can change the pace of each (for example, 11 September 2001 made governmental action quicker, while controversy over stem cells has delayed the technology).

9.3 EVOLUTION: THE HARE RESTS FOR A WHILE

In the evolution phase (1975–1986) of GMO oversight, the hare (GMO technologies) did not get too far before she rested at two key junctures, shortly after the creation of the first GMOs and shortly before the release of the first GMOs into the environment (see points 1 and 2 in Figure 9. 1). The tortoise meanwhile had a chance to catch up. The first pause in the race was prompted by the Asilomar conference, which brought together scientists and the media to discuss whether experiments with recombinant DNA (rDNA) in the laboratory warranted precaution and put some consensus restrictions on GM technology.[6] It was unique to the evolution phase of GMO oversight, as it was a voluntary pause on the part of the hare, marking the first time GM technologies stopped on their own volition (Figure 9.2a). This Asilomar meeting eventually led to the involvement of National Institutes of Health (NIH) Recombinant DNA Advisory Committee (RAC), which was then tasked with oversight of laboratory experiments involving GMOs.[7]

The second point where the hare paused during the evolution phase, she had to do so. Over time, researchers wanted to move GMOs out of the laboratory and into the environment and marketplace. The Coordinated Framework for the Regulation of Biotechnology (CFRB) was formulated in 1986[8] in response to congressional hearings and court cases over the release of the ice-minus GM bacteria into the environment. CFRB instructed three federal agencies, the US Environmental Protection Agency (EPA), the US Food and Drug Administration (FDA) and the US Department of Agriculture (USDA) to use the Toxic Substances Control Act (TSCA), Federal Insecticide, Fungicide, and Rodenticide Act

---

[6] Paul Berg, David Baltimore, S Brenner, R. O. Roblin and Maxine F. Singer, "Summary Statement of the Asilomar Conference on Recombinant DNA Molecules," 72 PROC. NAT'L ACAD. SCI. U.S. AM. 1981 (1975).

[7] Bernard Talbot, "Introduction to Recombinant DNA Research, Development and Evolution of the NIH Guidelines, and Proposed Legislation," 12 U. TOL. L. REV. 804 (1980).

[8] Coordinated Framework for Regulation of Biotechnology; Announcement of Policy; Notice for Public Comment, 51 Fed. Reg. 23302 (June 26, 1986).

Evolution (1970--1986)

- (1960-70s) Scientists develop recombinant DNA techniques and eventually are able to synthesize and replicate genes resulting in the ability to produce human growth hormone.
- (1974-75) The leading genetic scientists convene at the Asilomar meeting and encourage the U.S. Govt. to develop guidelines for regulating experiment using rDNA.
- (1976) U.S. Govt. sets up the Recombinant DNA Advisory Committee (RAC) through the National Institutes of Health (NIH 1978).
- (1980) Laboratory applications (such as the development of a gene-synthesizing machine) and industrial applications (such as Exxon's oil-eating microorganism) were emerging.
- (1982) RAC reviews its 1st proposed environmental release – the introduction of ice minus during the process of field trials.
- (1984-86) Legal controversy emerges involving the "ice minus" bacterium and it's release into the environment (USC 1983).
- (1986) The Office of Science and Technology Policy (OSTP) forms a working group, called the Biotechnology Safety Coordinating Committee (BSCC), and develops a three agency jurisdictional model known as the Coordinated Framework for the Regulation of Biotechnology (CFRB)

Implementation (1986--2002)

- (1987-1994) USDA, FDA, and EPA interpret existing laws (such as the Federal Insecticide, Fungicide, and Rodenticide Act (FIFRA), the Toxic Substance Control Act (TOSCA), and the Federal Plant and Pesticide Act (FPPA)) and promulgates regulations ad policies regulations concerning GEOs.
- (1992) FDA issues novel foods policy which states that foods derived from new plant varieties do not substantially differ from conventional counterparts.
- (1993) USDA issues regulatory guidelines for how a plant would become commercialized, or brought to the market.
- (1994) EPA develops rules for plant incorporated protectants (PIPs).
- (1994-95) The first commercial GEO products, such as the Flavr-Savr tomato, emerge on the market.
- (1999) A study raises concerns about possible negative effects of Bt Corn on Monarch butterflies, however, subsequent field research has shown little possibility of harm.
- (2000) StarLink, a form of Bt Corn that is thought to cause allergic reactions in small numbers of people and therefore not allowed in the food supply, is found in food supply causing recalls of some food and controversy over biotechnology regulation.
- (2000) EPA conducts a meeting of its Scientific Advisory Panel (SAP) and considers re-registering Bt corn.
- (2000) The National Research Council (NRC) publishes a report on EPAs review of pest-protected GE plants. FDA asks for comments on a mandatory pre-market safety review of GE foods.
- (2000-01) The Office of Science and Technology Policy (OSTP) and the Council on Environmental Quality (CEQ) conducts 6 case studies on the federal regulation of agricultural biotechnology
- (2001) The FDA publishes draft labeling guidance for companies that wish to label their foods as containing, or not containing, genetically modified ingredients and for making the agency's review process mandatory (FDA 2001). The process is still not mandatory (2008).
- (2001) The EPA publishes its finalized rules for the regulation of PIPs.
- (2002) The NRC publishes a report stating that the USDA should more rigorously review the potential environmental effects that transgenic plants may cause before approval for commercial use
- (2002) Prodigene Inc. was fined for tainting soybeans in Nebraska with experimental corn used for producing human vaccines.

a

b

180

Adaptation (2002–2010)

- (2002) OSTP publishes a notice, in the federal register, proposing actions to update field test requirements for industrial plants, although it concerns mainly harvesting and extraction.
- (2002) Government Accounting Office calls for enhancement of FDA's voluntary evaluation process for GE foods.
- (2003) USDA introduces enhanced safety guidelines, including more stringent confinement measures, for field trials of pharmaceutical and industrial GE plants - creating the permitting process interim rule.
- (2004) The NRC and the Institute of Medicine (IOM) publishes a report about the unintended consequences of GE foods in which they recommend that greater scrutiny be given to foods containing new compounds (
- (2004) The FDA issues draft guidance for industry regarding the Early Food Safety Evaluation of new non-pesticidal proteins produced by new plant varieties intended for food use (2005) USDA introduces finalized safety guidelines for field trials on industrial and pharmaceutical GE plants - creating the permitting process Final rule.
- (2005) Syngenta's Bt-10 corn is found in food supply, however FDA deems it no risk and therefore not illegal (although planting Bt-10 corn in the U.S. is illegal). Field trials of Ht Creeping Bentgrass contaminate wild grass varieties in Oregon.
- (2006) FDA finalizes its guidelines to industry regarding Early Food Safety Evaluation.
- (2006) Bayer notifies FDA and USDA that trace amounts of LLRICE601 (a form of Ht rice) has been detected in commercial rice varieties and may have entered the food supply. The USDA then determines that LLRICE601 does not pose a threat based on available data (provided by Bayer) and deregulates it.
- (2007) The Animal Plant Health Inspection Service (APHIS) publishes its first Environmental Impact Statement (EIS) in compliance with the National Environmental Protection Act (NEPA) concerning its regulation of GEOs. This occurred partially in response to lawsuits challenging USDA for improperly regulating GE plants.
- (2008) Unapproved GE corn seed inadvertently sold to farmers by Dow Agrosciences affiliate Mycogen Seeds
- (2008-2009) USDA proposes revisions to its regulation under the Plant Pest Act sparking 15,000 comments and public meetings.
- (2009-2010) In separate cases filed against USDA, federal judges revoke USDA's environmental assessment (EA) and deregulation of herbicide tolerant (HT) alfalfa and HT sugar beets. The judges rule that the EAs inadequately addressed the risks posed by genetically engineered crops, specifically regarding gene flow and the possible contamination of organic crop varieties. USDA was ordered to complete the more intensive environmental impact statement (EIS) for each crop.

C

Figure 9.2 Initial Phases of GMOs Oversight Development

181

(FIFRA), Federal Food, Drug, and Cosmetic Act (FFDCA) and the Federal Plant Pest Act (FPPA) to regulate the products of biotechnology and GMOs. In addition, the National Environmental Policy Act (NEPA) requires agencies to prepare detailed assessment of the impacts on the human environment.[9] Agencies can also prepare a more limited environmental assessment (EA) under NEPA if they are not sure whether the impacts are "significant." The CFRB relied on the policies that the product, not process, should be the focus of regulation and that no new laws were needed to cover GMOs and GMO products. It was set up as a science-based decision-making system, with additional focus on utilitarian accountings of costs and benefits in accordance with broader US regulatory policy.

The creation of the CFRB marks the end of the evolution phase, and GMO technologies were now poised to take off with an oversight system in place. In a temporal sense, the oversight system was well paced with the technology during this phase (Figure 9.1). Oversight systems were put in place as the technologies matured (Asilomar for laboratory work) or before they were deployed (CFRB for environmental release of GM bacteria and plants). This phase can be thought of as a pacing friendly phase, in which the tortoise and hare are aware of each other's progress and are somewhat synchronized. Because the template of the CFRB developed through existing laws, formulating new statutes did not delay governance. The framework was meant to be flexible, allowing for agency interpretation of existing laws to address health and environmental concerns. This phase can be described as pacing through coordinated interagency policy making.

However, pacing in a temporal sense did not seem to be enough to make a difference in the eyes of the public. Prominent civil society groups opposed the CFRB.[10] Many opponents believed that new and focused policies and laws were needed to fully cover the risks and societal impacts associated with GMOs and their products. They argued that biotechnology is a process that presents new risks and requires special regulation. This viewpoint runs counter to the US policy of focusing on these products being the same in kind as those that are bred by conventional means.[11] Twenty years later, this framework is still

---

[9]   42 U.S.C. § 4332(2)(C) (2011).

[10]   George Gaskell, BIOTECHNOLOGY, 1996–2000: THE YEARS OF CONTROVERSY (2001).

[11]   National Academy Of Sciences, INTRODUCTION OF RECOMBINANT DNA-ENGINEERED ORGANISMS INTO THE ENVIRONMENT: KEY ISSUES (1987).

operational, although it has evolved over time (Figure 9.2). Scholars look back now on this phase as a fairly closed-door elite process for making initial decisions about GMOs. Asilomar was limited largely to scientists and the media, and the White House Office of Science and Technology Policy's formulation of the CFRB was insulated from public discussion. The CFRB's operations did not develop mechanisms to engage stakeholders and the public. The science-based premise of the CFRB left little to no room for discussion about values like preserving nature, minimal inputs to agriculture, thinking about unintended effects, and fairness of risk and benefit distributions.[12] Furthermore, the system was set up as very flexible in its loose interpretation of existing laws (for example, GM microbes as toxic chemicals),[13] which as time went on, proved to be susceptible to changing political, economic and social winds. The story begins to unfold with problems in oversight beyond temporal pacing.

9.4 IMPLEMENTATION: THE HARE TAKES OFF

General and flexible laws designated under the CFRB needed to be more concretely interpreted to regulate GMOs. The interpretation of the CFRB and the explosion of GM products in agriculture mark the second phase of oversight: implementation (Figure 9.2b). During this phase, the boundaries of various statutes significantly stretched to promulgate agency regulations for diverse products. GM plants were regulated as plant pests under FPPA, because they often contained engineered sequences from viruses and bacteria that cause plant disease or because the plants themselves can be considered plant pests.[14] The Animal Plant Health and Inspection Service (APHIS) would be the lead USDA agency for biotech crops. Under FIFRA and FFDCA, EPA regulated GM plants engineered with pesticidal-like proteins (plant incorporated protectants).[15] Inter-generic GM microorganisms were regulated as toxic chemicals

---

[12] Paul B. Thompson, FOOD BIOTECHNOLOGY IN ETHICAL PERSPECTIVE (2nd edn. 2007).

[13] Kuzma et al., supra note 3.

[14] Introduction of Organisms and Products Altered or Produced Through Genetic Engineering Which are Plant Pests or Which There is Reason to Believe are Plant Pests, 7 C.F.R. § 340 (1987, 1997); Genetically Engineered Organisms and Products: Notification Procedures for the Introduction of Certain Regulated Articles and Petition for Nonregulated Status, 58 Fed. Reg. 17044 (March 31, 1993).

[15] Regulations under the Federal Fungicide, Insecticide, and Rodenticide Act for Plant-Incorporated Protectants, 66 Fed. Reg. 37,855 (July 19, 2001).

under TSCA.[16] Under FFDCA, the FDA reviewed GM or bioengineered foods through a voluntary consultation mechanism.[17]

Although the CFRB was established at the beginning of implementation, the rules and guidance documents under it did not take shape until later. The technology blossomed while the roles of some agencies lagged. For example, the EPA's Plant Incorporated Protectant (PIP) rule was proposed in 1994 but not finalized until 2001, although companies complied with the draft rule in the interim. On the flip side, USDA had its field trial regulations in place in 1987, before crop field trials took place, and FDA published its guidance for novel foods produced through biotechnology in 1992, before the first GM foods entered the market. During the early part of the implementation phase, regulations kept pace with the first generation of GM products, namely genetically engineered microbes (GEMs), herbicide tolerant (Ht) and pest-resistant (primarily Bt) crops. The hare and the tortoise were moving at a similar pace (Figure 9.1, point 2). The system paced with technology through formal regulations and policies (pacing through regulation). GM crops, especially major commodity crops with insect-resistant (primarily Bt) and herbicide-tolerant (Ht) engineered genes, exploded onto the market while thousands of field trials were conducted.

Toward the end of the implementation phase – with new findings and technological developments – regulations, guidances and policies stretched in place. The CFRB and oversight system were not prepared to deal with pharmaceutical or industrial chemical production in plants, GM fish or GM insects. The hare (technology) took off ahead of the tortoise (oversight system) (Figure 9.1, point 3). Then credibility was shaken by reports of harm to Monarch butterflies from Bt pollen; comingling of Starlink Bt corn, which was not approved for human food due to concerns about allergenicity, with corn destined for human food; the detection of gene flow between Bt and non-Bt maize in Mexico; and contamination of soybeans with corn containing engineered pharmaceuticals.[18] For example, the company ProdiGene failed to eliminate volunteer GM corn plants producing pharmaceuticals from a soybean

---

[16]   40 C.F.R. § 725 (1997).
[17]   Statement of Policy: Foods Derived from New Plant Varieties, 57 Fed. Reg. 22,984 (May 29, 1992).
[18]   CONG. RESEARCH SERV., RS20732, STARLINK™ CORN CONTROVERSY: BACKGROUND (2001); J.L. Fox, Puzzling Industry Response to Prodigene Fiasco, 21 NATURE BIOTECHNOLOGY 3 (2003); D. Quist and I. Chapela, Transgenic DNA Introgressed into Traditional Maize Landraces in Oaxaca, Mexico, 414 NATURE 541 (2001).

crop planted later in the same field and destined for human food.[19] USDA imposed a fine of $250 000, and ProdiGene had to reimburse the federal government $3 million for the destruction of the contaminated soybeans. This and other controversies mark the end of the implementation phase (Figure 9.2b).

The above controversies may have taken a toll on the production and marketing of GM crops with novel traits, as well as non-commodity GM crops (minor and orphan crops), because the diversity and number of approvals for full USDA deregulation decreased in 2000 (Figure 9.3).[20] There were also media reports of large companies pulling out of certain GM crop markets (for example, pharmaceutical production in crops).[21] Toward the end of the implementation phase, the hare slowed down a bit, while regulators increased their attention to risks and their management.

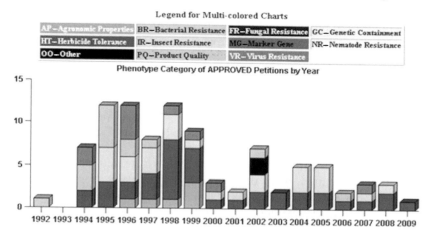

Figure 9.3   Approved Petitions for Deregulation by USDA Per Year

---

[19]   Fox, supra note 18.

[20]   Info. Sys. Biotechnology, Crops No Longer Regulated By USDA (Petitions for De-Regulation), VIRGINIA TECH (2012), available at http://www.isb.vt.edu/search-petition-data.aspx.

[21]   P. Byrne, Bio-pharming, COLORADO STATE UNIVERSITY EXTENSION (April 2008), available at http://www.ext.colostate.edu/pubs/crops/00307.html.

9.5   ADAPTATION: THE TORTOISE CHUGS ALONG TO CATCH THE HARE

The adaptation phase is marked by changes to the CFRB and its operation in order to mitigate risks arising from the implementation phase and to expand the scope of the products covered by the CFRB. Federal agencies involved in US oversight did not promulgate new regulations to deal with emerging concerns, but rather chose to deal with them through regulatory guidances. Key areas of concern were health risks such as allergenicity, co-mingling of unapproved and approved GM varieties, and gene flow to wild relatives or neighboring crops. For example, the National Organic Program rule of 2000 prohibited the intentional use of GM crops in certified organic foods.[22] Troubles began with the coexistence of organic and GM crops at the end of the implementation phase. Organic farmers in the United States became increasingly concerned that GM crops would cross-pollinate with their crops, and they would no longer be able to certify their products as organic.

Although the laws and interpretations for GM crops largely remained the same during this phase, several guidance documents and regulatory policies were published to address emerging GM plant products and their impacts (Figure 9.1, points 3 and 4, and Figure 9.2c). Public and stakeholder reactions to new risk information and perceived failures of the system prompted these adaptations. The tortoise and hare took turns resting and running, with the tortoise employing a pacing through guidance. For example, FDA put out a guidance document to improve early food-safety evaluation of GM crops with non-pesticidal proteins,[23] and USDA published enhanced biosafety guidelines, including requiring more stringent confinement measures for growing GM plants containing pharmaceutical and industrial engineered proteins.[24] So far, the hare was not too far ahead of the tortoise with regard to GM crops.

---

[22]   National Organic Program; Final Rule, 65 Fed. Reg. 80,548 (December 21, 2000).
[23]   US Food & Drug Administration, GUIDANCE FOR INDUSTRY: RECOMMENDATIONS FOR THE EARLY FOOD SAFETY EVALUATION OF NEW NON-PESTICIDAL PROTEINS PRODUCED BY NEW PLANT VARIETIES INTENDED FOR FOOD USE (June 2006), available at http://www.fda.gov/downloads/AnimalVeterinary/GuidanceComplianceEnforcement/GuidanceforIndustry/UCM113903.pdf (accessed 17 July, 2013).
[24]   Field Testing of Plants Engineered to Produce Pharmaceutical and Industrial Compounds, 68 Fed. Reg. 11,337 (March 10, 2003).

However, for GM animals during this phase, the hare was stalled. At the time of development and during initial phases of implementation, the CFRB framework did not specifically consider the regulation of GM insects, trees, plant pharmaceuticals, fish or mammals. There is still some ambiguity about oversight for these GMOs and their products. The first GM animal for food (transgenic) use had been waiting for approval for over a decade, in part because of a lack of a regulatory guidance to interpret existing laws. Then, in 2009, the FDA proposed to oversee GM animals as "investigational new animal drugs" under FFDCA.[25] However, the first test case of growth-enhanced transgenic salmon for human consumption is still waiting approval due to controversies surrounding its use. FDA did approve a transgenic goat producing the human drug antithrombin in its milk,[26] but not for food use. GM animals seem to be resting in the tortoise's hands.

Toward the end of this phase, the hare was also stalled with regard to GM crops, allowing the tortoise to catch up. An outside racer, the NGO community, filed legal suits against the industry and USDA for lack of compliance with NEPA. These court battles distracted industry and caused great uncertainty about the future of GM crop approval processes through USDA. USDA was forced to complete its first ever Environmental Impact Statements (EISs) for GM crop deregulation (market approval) for Ht alfalfa and Ht sugar beets.[27] Plantings were delayed, and it looked as though future pre-market decisions on particular GM crops would likely be slowed with more intensive NEPA compliance requirements. The court cases are discussed further in the revolution phase, as they mark the end of adaptation and beginning of changes in US policies for governance.

The final part of this phase includes the tortoise's attempt to revise its regulations through a formal rulemaking process. In a draft rule, USDA tried to revise and clarify its authority for GM crops under the newer Plant Protection Act of 2000 (PPA) (replacing the FPPA in the CFRB) by applying its noxious weed authorities to GM crops and changing the

---

[25]   US Food & Drug Administration, GUIDANCE FOR INDUSTRY: REGULATION OF GENETICALLY ENGINEERED ANIMALS CONTAINING HERITABLE RDNA CONSTRUCTS (January 15, 2009), available at http://www.fda.gov/AnimalVeterinary/GuidanceComplianceEnforcement/Guidancefor Industry /ucm123631.htm (accessed 17 July 2013).
[26]   Id.
[27]   Congressional Research Service, R41395, DEREGULATING GENETICALLY ENGINEERED ALFALFA AND SUGAR BEETS: LEGAL AND ADMINISTRATIVE RESPONSES (September 10, 2012).

notification or permit process to a tiered permit system based on categories of risk, including a conditionally exempt category.[28] An EIS on the draft rule was also published. Because technological advances have led to the possibility of developing GM crops (or other GMOs) that do not fit within the plant pest definition but still might cause harm, the proposed regulations would also subject GMOs to oversight based on known plant-pest and noxious-weed risks of the parent organisms, or the traits of the organism, or the possibility of unknown risks as a plant pest or noxious weed when insufficient information is available. However, USDA never finalized the 2009 proposed rule. The tortoise's attempt to move with the technology was thwarted.

In summary, during the adaptation phase, pacing through guidance worked until the external stakeholder pressure and legal battles became too much for the tortoise and hare. They needed to use a different tactic.

9.6   REVOLUTION: THE HARE IS TIED UP, THEN TAKES OFF WITH TORTOISE'S HELP

The choice in changing race tactics marks the beginning of the revolution phase. During this phase, the tortoise paced through fundamental shifts in policy, including a change in the interpretation of the laws and exertion of authorities on which the CFRB was based (Figure 9.4). This change occurred following court decisions related to GM alfalfa. The GM alfalfa story is outlined below to provide background on some of the key factors that likely changed GM crop governance policies at USDA.

In 2005, USDA approved Monsanto's Roundup Ready herbicide-tolerant (RR, Ht) alfalfa for commercial planting and decided that the crop did not pose a plant pest risk. Under NEPA, instead of preparing a full EIS for this decision, the agency issued an EA-FONSI (Environmental Assessment, Finding of No Significant Impact). Organic farmers and NGOs challenged the adequacy of USDA's EA on RR alfalfa in federal district court, arguing that the agency should have prepared a full EIS. Until then, USDA had never prepared a full EIS for any GM crop approval. In 2007, the district court sided with the plaintiffs and prohibited Monsanto from planting or selling RR alfalfa seeds until a full

---

[28]   Importation, Interstate Movement, and Release into the Environment of Certain Genetically Engineered Organisms: Proposed Rule, 73 Fed. Reg. 60,008 (October 9, 2008).

Revolution (2010–present)

- (2010) USDA decides not to exert authority for Zinc Finger Nuclease low phytate corn
- (2011) In January, Congress has hearing about GE alfalfa case. Several members of Congress question USDA's authority under the PPA to regulate GM crops at all.
- (2011) After completing the HT alfalfa EIS, USDA decides to fully deregulate HT alfalfa allowing for its unrestricted use.
- (2011) While in the process of completing the EIS for HT sugar beets, USDA partially deregulates them allowing for their restricted commercial use
- (2011) USDA approves amylase corn without EIS
- (2011-2012) USDA deregulates several GE crops without EIS

Figure 9.4   Present Phase of GM Crop Oversight

EIS was completed,[29] putting an injunction on further planting of GM alfalfa in place. It is estimated that approximately 5500 growers across 263 000 acres planted GM alfalfa before the injunction, and most of these farmers were planning on planting GM alfalfa seeds in the upcoming growing season. With all these growers in jeopardy, Monsanto appealed to the Ninth Circuit, but the court upheld the ban.[30] Monsanto then appealed to the US Supreme Court, and at that point, USDA decided not to join the case and continued to work on a final EIS for Ht alfalfa. The Supreme Court decided to take up the case, *Monsanto v. Geertson Seed Farms*,[31] and granted Monsanto's petition that the court review the scope of the permanent injunction against further planting of Ht alfalfa.

The main environmental issue in the Ht RR alfalfa case was cross-pollination with weedy relatives or organic crops. Harm could occur to the organic industry if RR alfalfa contaminated neighboring organic crops. The presence of GM alfalfa could cause organic farmers to lose their market shares. Coexistence of organic with GM crops became a focal point of the policy conversation. Scientific concerns were focused

---

[29]   Geertson Farms, Inc. v. Johanns, No. 06-01075, 2007 WL 776146 (N.D. Cal. March 12, 2007) (preliminary injunction); Geertson Farms, Inc. v. Johanns, 2007 WL 1302981 (N.D. Cal. May 3, 2007) (permanent injunction); Geertson Farms, Inc. v. Johanns, 2007 WL 1302981, *5 (N.D. Cal. May 3, 2007).

[30]   Geertson Seed Farms v. Johanns, 570 F.3d 1130 (9th Cir. 2009), rev'd sub nom. Monsanto Co. v. Geertson Seed Farms, 130 S. Ct. 2743 (2010).

[31]   Monsanto Co. v. Geertson Seed Farms, 130 S. Ct. 2743 (2010).

on the possibility that wild relatives of alfalfa could be contaminated and become super weeds, resistant to Roundup or other herbicides.

In June 2010, the Supreme Court ruled that the permanent injunction against planting or selling should be removed. It noted that the district court had overreached procedurally in halting the plantings, and as a result, Ht RR alfalfa could be planted while the USDA completed the EIS. However, the Supreme Court justices did not remove the need for an EIS, and they stressed an intermediate option of partial deregulation (planting with geographic restrictions). The court's ruling also suggested that environmental harm included economic effects such as reduced agricultural yield or loss of market due to genetic contamination.[32] Thus, the ruling was interpreted as a victory by both sides. RR alfalfa could be planted in the spring, but the EIS could consider economic harm. The court suggested partial deregulation as an option, although ultimately leaving the decision to USDA.

In December 2010, USDA published the final EIS, outlining three options in the document: (1) ban the commercial planting of RR GM alfalfa (no deregulation); (2) approve it with planting restrictions such as greater isolation distances from other crops (partial deregulation); or (3) approve it with no planting restrictions (full deregulation).[33] USDA indicated that it was seriously considering the latter two options. In the document, USDA argued that "[b]ecause Congress has mandated a science-based approach in APHIS regulations and because there is no basis in science for banning the release of GT alfalfa, a blanket prohibition of the release of GT alfalfa would contravene Congressional intent and must be rejected."[34] The USDA chose in this statement to privilege the natural science over socioeconomic harms.

After the EIS was published, both sides put pressure on USDA regarding its impending decision whether to partially or fully deregulate RR alfalfa. The organic industry and NGO community were arguing for no deregulation or partial deregulation. National agricultural commodity associations lobbied against partial deregulation, arguing that it would be a significant departure from regulatory practices and could have negative

---

[32]   Congressional Research Service, supra note 27.

[33]   US Department of Agriculture, GLYPHOSPHATE-TOLERANT ALFALFA EVENTS J101 AND J163: REQUEST FOR NONREGULATED STATUS FINAL ENVIRONMENTAL IMPACT STATEMENT (December 2010), available at http://www.aphis.usda.gov/biotechnology/alfalfa_documents.shtml (accessed 17 July 2013).

[34]   Id. at 14.

impacts on current trade agreements.[35] Soon thereafter, the House Committee on Agriculture hosted a forum on January 20, 2011to discuss agricultural biotechnology regulation and the GM alfalfa situation with USDA Secretary Tom Vilsack. House members intensely questioned the secretary about the partial deregulation option, and the vast majority of members indicated their support for full deregulation. The members also more broadly challenged USDA's authority for regulating GM crops under the PPA at all.[36]

On 27 January 2011, Vilsack announced that USDA was granting GM alfalfa full deregulation on the basis that it posed no greater plant pest risk than other conventional alfalfa varieties and that any option other than full deregulation was inconsistent with their regulatory authority under PPA. USDA decided that it would have no further control over the planting and distribution of GM alfalfa.

As a result of the USDA's alfalfa decision, the doors opened for other GM crops to enter the market. Responding to congressional pressure, USDA backed off its authority under the PPA and NEPA. For example, a new GM crop, Syngenta's amylase corn, was approved for market release. Amylase corn is engineered to produce a thermostable version of the enzyme alpha-amylase that breaks down starch into sugar for ethanol production. USDA deregulated it in February 2011, a month after GM alfalfa, without a full EIS. Several stakeholder groups including food producers believe that USDA failed to adequately consider the impact of GM amylase corn on human health, the environment or the livelihood of farmers.[37] In the past two years, approvals of new GM crops have spiked.

Since the alfalfa case, USDA has also, quite radically, decided that several GM crops do not fall under its plant pest authority at all, including crops produced through new targeted genetic modification methods (for example, Dow's low phytate corn)[38] and Ht Kentucky bluegrass. Initial biotechnology techniques used sequences from plant pests in the engineering process. In 2000, USDA stated that the plant pest designation was just a hook to regulate all GM crops under existing laws and that the agency would use this hook to regulate GM crops regardless

---

[35] Congressional Research Service supra note 27.

[36] Forum to Review the Biotechnology Product Regulatory Approval Process: A Forum Before the H. Comm. on Agric., 112th Cong. (January 20, 2011).

[37] Emily Waltz, "Amylase Corn Sparks Worries," 29 NATURE BIOTECH-NOLOGY 1063 (2011).

[38] J. Kuzma and A. Kokotovich, "Renegotiating GM Crop Regulation," 12 EMBO REPORTS 883 (2011); Emily Waltz, "Tiptoeing Around Transgenics," 30 NATURE BIOTECHNOLOGY 215 (2012).

of whether crops contained plant pest sequences.[39] A decade later, with newer genetic engineering methods, these pest sequences are no longer needed for engineering, and USDA has recently decided not to use PPA for GM crops not containing plant pest sequences. USDA is now choosing to interpret its authority under the PPA strictly, even though under 7 CFR part 340 the administrator of APHIS has the ability to declare a GM organism a regulated article if:

> it has been genetically engineered from a donor organism, recipient organism, or vector or vector agent listed in 340.2 and the listed organism meets the definition of "plant pest" or is an unclassified organism and/or an organism whose classification is unknown, *or if the Administrator determines that the GM organism is a plant pest or has reason to believe it is a plant pest.*[40]

Thus, its new interpretation is a revolutionary shift. Recently Ht RR Kentucky bluegrass, which does not contain plant pest sequences (but could have been designated as a plant pest if the administrator wanted to do so) was exempted from USDA review under the PPA. Free planting of Ht bluegrass is allowed without any formal regulatory review, despite concerns about gene flow to neighboring grasses and increased weed resistance to Roundup.[41] USDA did not consider the risks through the permit or deregulation process and did not require any assessments for market release.

This period of contested legal disputes and subsequent USDA policy shifts can be interpreted as a revolution in interpretation of USDA authority. Continued legal suits would likely have pressured USDA to complete a rigorous EIS for every GM crop it considered under PPA. Each EIS would be a vast undertaking, likely taking years to complete. Congress questioned USDA authority more broadly, and most members would prefer a rapid approval of GM crops. Therefore, it has become more feasible and politically expedient for USDA not to exert authority under the PPA for GM crops that are not explicitly containing plant pest sequences or not clearly plant pests. USDA took the road that was more feasible given current resources and the political climate against tight regulation.

---

[39]   National Research Council, GENETICALLY MODIFIED PEST-PROTECTED PLANTS: SCIENCE AND REGULATION (2000).

[40]   7 CFR part 340 (emphasis added).

[41]   Paul Voosen, "In Major Shift, USDA Clears Way for Modified Bluegrass," NYTIMES.COM, July 6, 2011, available at http://www.nytimes.com/gwire/2011/07/06/06greenwire-in-major-shift-usda-clears-way-for-modified-bl-51693.html (accessed 17 July 2013).

USDA is trying to expedite its review process even for the GM plants captured under the PPA. USDA estimates that the cost of a draft EA generally ranges from $60 000 to $80 000, and the cost of a complete EIS can be over $1 million.[42] A new model proposed recently in the Federal Register would allow independent contractors to prepare EAs or EISs, funded by a cooperative agreement between the petitioner and APHIS. USDA is currently in the process of implementing the pilot program. Many groups are concerned that this model allows producers of biotech crops even more control over the NEPA process by allowing them or other private contractors to author NEPA assessment documents.[43]

In summary, this shift to pacing through policy marks the revolution phase. Initially USDA could not keep up with legal suits and the need to produce EISs, and delays resulted (Figure 9.1, point 4). Delays were likely to continue from this contested process, so USDA decided to forgo the regulation of GM crops that are not squarely plant pests. Some GM crops and products will still take a while to review, but EISs will not be the norm (for example, amylase corn), and many products will not be captured under the PPA at all (such as Ht bentgrass and low phytate GM corn). In terms of the fable, the hare caught the tailwinds of an anti-regulatory climate, and the tortoise decided to give up a leg of the race (Figure 9.1, point 5).

9.7 PROPER PACING

The US GMO oversight system formed and adjusted alongside the progress of GM crops, although it did lag behind at times (Figure 9.1). It paced in a temporal sense in different ways in the four phases: pacing through coordinated interagency policy making; pacing through regulations; pacing through guidance; and pacing through policy shift (that is, reinterpretation of authorities). In comparison to other regulatory systems – such as the one still evolving for nanotechnology – the GMO oversight

---

[42] Solicitation of Letters of Interest to Participate in National Environmental Policy Act Pilot Project, 76 Fed. Reg. 19,310 (April 7, 2011).
[43] E. Burkett, "Who Should Conduct Biotech Crop Assessments?" FOOD SAFETY NEWS, April 25, 2011, available at http://www.foodsafetynews.com/2011/04/look-who-is-going-to-be-doing-the-environmental-assessments/ (accessed 17 July 2013).

system for crops was arguably proactive in its beginnings.[44] For example, the NIH guidelines for laboratory use were in place shortly after rDNA techniques arose, the CFRB was in place before the technology was deployed in the field, and most agency regulations were promulgated before GM crops hit the market. The system responded to emerging conditions and external events, technological development, risk concerns and stakeholder pressure.

In general, government decisions for oversight of GM crops seemed to pace well with the technology in a temporal sense. However, many stakeholders do not see the GMO system as appropriate in the longer term for ensuring health and environmental safety.[45] We have found in previous work that stakeholders and experts familiar with the GMO oversight system rate it as highly flexible, but with weak legal grounding, few opportunities for public input, poor treatment of uncertainty, low transparency, little post-market monitoring and a lack of information for consumer choice.[46] The high flexibility has allowed the agencies to adapt with changes in technology, but it also allowed for changes in regulatory policy based on political winds instead of risk assessments (for example, Ht bluegrass).

GM crops could be considered a market success in some ways, like market penetration and industry growth, but a public failure from a science and technology policy perspective.[47] Public failures can stem from many factors, including inadequate policies that are incongruent with public values and consider only short time-horizons. The increasing number of products on the market labeled as GM-free and more widespread pressure requiring labeling of GM foods (for example, Proposition 37 in California) suggests that GM crops are a public failure.[48] NGOs and organic farming groups are not pleased with the system and continue to pursue legal challenges against the planting of

---

[44]   Jennifer Kuzma, Pouya Najmaie and Joel Larson, "Evaluating Oversight Systems for Emerging Technologies: A Case Study of Genetically Engineered Organisms," 37 J.L. MED. & ETHICS 546 (2009); Gurumurthy Ramachandran et al., "Recommendations for Oversight of Nanobiotechnology: Dynamic Oversight for Complex and Convergent Technology," 13 J. NANOPARTICLE RES. 1345 (2011).

[45]   Kuzma et al., supra note 44; Thompson, supra note 122; Gaskell, supra note 10.

[46]   Kuzma et al., supra note 44.

[47]   B. Bozeman and D. Sarewitz, "Public Values and Public Failure in US Science Policy," 32 SCI. & PUBLIC POL'Y 119 (2005).

[48]   Id.

other crops such as Ht GM canola in Oregon,[49] Ht RR Kentucky blue grass[50] and GM eucalyptus trees.[51] Controversies are being played out in the new farm bill negotiations over revising regulatory processes as well.[52]

Based on the GM crop story, a broader notion of pacing that incorporates not only elements of time but also includes responsiveness to potential ecological and social harms, as well as public desires for information, transparency and voice, is suggested. I term this "proper pacing." The oversight decisions for emerging technologies like GMOs, nanotechnology, robotics, synthetic biology, geoengineering and neuro-technology will continue to be controversial. Stakeholders and the public – with different ideas and values about the role of technology in society, assurances of safety and equal sharing of risks and benefits – are currently forced to express their views outside the oversight system, looking in only through the public comment and rulemaking process or by challenging industry and agencies in court. On the flip side, industry and government are crippled by these challenges, delaying potentially beneficial and more acceptable technologies to flourish. Fundamental change in oversight is needed to address the proper pacing problem as we enter a world changing from the confluence of multiple new technologies with uncertain impacts. Innovation in oversight is needed to match technological innovation. But most importantly, this innovation in over-sight should create opportunities for dialogue and debate much earlier upstream in the technology decision making pipeline. It should be flexible in response to technologies, but the need for flexibility and the shape it takes should be decided with the input of multiple stakeholders.

Proper pacing could draw upon ideas that several scholars have proposed for new governance regimes. For example, in Marchant et al. (2011), the use of voluntary environmental programs, codes of conduct, anticipatory governance and administrative law tools are described for

---

[49]   "Planting of GM Canola Stayed in Oregon," GM WATCH NEWS, August 18, 2012, available at http://www.gmwatch.org/latest-listing/51-2012/14131-planting-of-gm-canola-stayed-in-oregon (accessed July 17, 2013).

[50]   International Center for Technology Assessment and the Center for Food Safety: Noxious Weed Status of Kentucky Bluegrass Genetically Engineered for Herbicide Tolerance, 76 Fed. Reg. 39,8100 (July 7, 2011).

[51]   Emily Waltz, Cold-Tolerant Trees Win, 29 NATURE BIOTECHNOLOGY 1063 (2011).

[52]   Bob Meyer, "What About that "Biotech Rider" in the Farm Bill?," BROWNFIELD: AG NEWS FOR AMERICA, November 15, 2012, available at http://brownfieldagnews.com/2012/11/15/what-about-that-biotech-rider-in-the-farm-bill/ (accessed 17 July 2013).

better pacing.[53] These methods may be feasible in the short term for pacing. However, I argue that a more fundamental paradigm shift is needed for proper pacing. Putting aside the legal and political practicalities, I briefly describe one vision for proper pacing below.

9.8   ONE POSSIBLE VISION

For this vision, I draw upon three frameworks from our previous work: dynamic oversight; upstream oversight assessment (a subset of anticipatory governance); and strong objectivity. Dynamic oversight is based on the principles of inclusivity, reflexivity, anticipation and adaptation, with the ability to move oversight from hard (regulatory) to soft approaches (codes of conduct) and back again depending on new information. Given these principles and movement, we proposed a model where three chartered groups would form the heart of decision making:[54] (1) an interagency group; (2) a diverse stakeholder group; and (3) a wider public engagement coordinating group. For example, decision making authority would still rest with the agencies, but they would be publicly accountable to the input from the stakeholder group, much like a federal advisory committee. The public engagement process could be coordinated through science museums across the country and fed into the stakeholder and interagency group process. Communication among the three groups would be routine and iterative.

The dynamic oversight model would then be used for upstream oversight assessment, a subset of anticipatory governance focusing on particular products as cases for governance.[55] Case studies of products or technologies coming down the pipeline (for example, in early development) would be presented to this three-bodied system. The technologies and their potential impacts on society would be deliberated. The three bodies would raise gaps in legal authorities, risk assessment needs, and social and ethical concerns. The system would prepare for how to govern the particular applications in the future. For example, the interagency group could examine and clarify authorities; members of the stakeholder group would put policies or codes in place, conduct safety studies or continue the dialogue to ameliorate conflict in the future; and citizens

---

[53]   See supra note 2.

[54]   Ramachadran et al., supra note 44.

[55]   Jennifer Kuzma, James Romanchek and Adam Kokotovich, "Upstream Oversight Assessment for Agrifood Nanotechnology: A Case Studies Approach," 28 RISK ANALYSIS 1081 (2008).

who took part in the engagement could seek additional information about the technologies, voice their opinion to decision makers, and monitor their entry into the marketplace. Although this process would take time and resources, this upstream dialogue and deliberation is likely to be less costly than heated legal disputes and delays like in the case of GM RR alfalfa.

One important cornerstone that must be embedded in such a system is strong objectivity. With the position of strong objectivity, values shaping risk assessments and knowledge-based policy making are identified and critically evaluated, and an intermediate position between the idea that science is purely objective and the idea of relativism is taken.[56] Scientific practices and knowledge claims associated with oversight of emerging technologies would be scrutinized from diverse perspectives to increase the objectivity by maximizing to the extent possible the standpoints from which scientific and social impacts are assessed. The idea is that with dynamic oversight, upstream oversight assessment and strong objectivity, the oversight system becomes more legitimate and rigorous, properly pacing with technological advances by repeated and iterative discussions among the three groups in the system.

The GMO story tells us that change is needed. Other contemporary stories of nanotechnology, geoengineering, robotics and synthetic biology are repeating that lesson. I argue that this change should come through a paradigm shift. The model presented here may not be ideal, but it could be a starting point for discussion. The future depends on proper pacing to create a better relationship between society and emerging technologies.

---

[56]   Sandra Harding, "'Strong Objectivity': A Response to the New Objectivity Question," 104 SYNTHESE 331 (1995).

10. Innovative governance schemes for molecular diagnostics

Rachel A. Lindor and Gary E. Marchant

10.1 INTRODUCTION

Molecular diagnostics are expected to shepherd in a new paradigm in health care delivery known alternatively as personalized, individualized or precision medicine. The key distinction of this new approach is that health care decisions will be based on a patient's individual molecular profile rather than on population or group characteristics or averages. Thus, for example, the selection and dose of a drug will be based on the particular set of variants present in the patient's genome that affect drug metabolism. Diseases will be detected much earlier based on the presence or concentration of a protein biomarker in the patient's blood, rather than waiting for the disease to manifest in clinical symptoms. The prognosis and best treatment option for a cancer patient will be determined by the changes in gene expression patterns in the patient's tumor.

These new molecular-based strategies for predicting, diagnosing, and treating disease are all dependent on sophisticated new molecular diagnostic tests. These molecular diagnostic tests will be the quarterbacks of the new personalized medicine paradigm. Just as a quarterback in football surveys the situation on the field, selects the appropriate play, and then sets that play in motion, molecular diagnostics will play a central role in assessing a patient's health status and risks, selecting an appropriate intervention strategy, and then targeting the treatment based on the patient's molecular profile. As is the case with quarterbacks, molecular diagnostics will therefore be critical to the outcome, and at least when successful, will be high profile, sophisticated, and high value.

The challenge for personalized medicine is that this model for molecular diagnostics is inconsistent with traditional economic and regulatory assumptions about diagnostic products. Extending the football metaphor another step, diagnostic tests have until recently been more analogous to offensive linemen in football. They have been relatively low cost, are called upon to perform tasks that are relatively straightforward and simple, and are generally fungible. The question facing the regulatory system therefore is how to transform diagnostics from being the linemen to the quarterbacks of medicine.

The frameworks for the regulation and reimbursement of diagnostics, combined with outdated business models and provider practices, stand as barriers to the successful development and commercialization of sophisticated molecular diagnostics that are critical to personalized medicine. In this chapter, we first expand on how path dependence has resulted in outdated frameworks for molecular diagnostics, and how this is slowing the uptake of personalized medicine. We then describe two innovative new programs – coverage with evidence (CED) and parallel review – that may expedite the commercial development of molecular diagnostics. The chapter concludes with some comments on the implications of these governance innovations for the "pacing problem."[1]

10.2 REGULATORY AND REIMBURSEMENT CHALLENGES FOR MOLECULAR DIAGNOSTICS

The sophisticated new molecular diagnostic tests that are the foundation of personalized medicine are encountering significant challenges and uncertainties with regard to both regulation and reimbursement. These problems are summarized below.

10.2.1 Regulation of Molecular Diagnostic Tests

There are two potential regulatory pathways for molecular diagnostics: one through the Food and Drug Administration (FDA), and the other pursuant to the Clinical Laboratory Improvement Act (CLIA) administered by the Centers for Medicare and Medicaid Services (CMS). The limitations and uncertainties about the application of both of these existing pathways to new molecular diagnostic technologies have generated much confusion among product sponsors. These uncertainties make it difficult for researchers and developers to successfully bring new products to market and threaten to slow or block the implementation of personalized medicine.

---

[1]   THE GROWING GAP BETWEEN EMERGING TECHNOLOGIES AND LEGAL-ETHICAL OVERSIGHT (Gary E. Marchant, Braden R. Allenby and Joseph R. Herkert, eds, 2011).

10.2.1.1 FDA regulation

Molecular diagnostics are regulated by the FDA as medical devices. Among the wide range of medical devices evaluated by the FDA, molecular diagnostics fall into a category known as "*in vitro* diagnostics" (IVDs). These devices, which rely on measurements at the molecular level of samples removed from an individual's body, are distinguished from *in vivo* devices such as blood pressure cuffs and thermometers, which rely on measurements taken directly from an individual. Since 1976, the FDA has been charged with ensuring that all medical devices it evaluates, including IVDs, provide a "reasonable assurance of safety and effectiveness."[2] Because the FDA evaluates devices before they enter regular clinical use, the agency relies almost exclusively on pre-market data supplied by the products' sponsors to evaluate safety and effectiveness.

In order to devote more resources to products that pose the greatest risks and uncertainty without slowing market entry for less risky products, the FDA groups new devices into one of three categories based on their level of risk and subjects those categories to corresponding levels of scrutiny. Class I and II devices are considered low enough risk that compliance with general controls (for example, good manufacturing practices, product labeling) or special controls (for example, post-market surveillance, patient registries) is sufficient to provide a reasonable assurance of safety and effectiveness, generally without the need for clinical data.[3] Class III devices are higher-risk and warrant additional investigation to ensure their safety and effectiveness, traditionally through a pre-market approval (PMA) process requiring clinical trials.[4]

For devices that would otherwise fall within Class III, the FDA seeks to limit its reliance on the resource-intensive PMA process by allowing a new device to be marketed without a PMA if the product's sponsor is able to demonstrate that its product is "substantially equivalent" to existing devices.[5] This is known as a "pre-market notification" or 510(k) clearance. Pursuant to this pathway, devices can be legally marketed simply by demonstrating that they are "substantially equivalent" to

---

[2]   Medical Device Amendments, Pub. L. No. 94-295, 90 Stat. 539 (1976).
[3]   21 CFR 860.3(c)(1–2).
[4]   21 CFR 860.3(c)(3).
[5]   Jeffrey Gibbs, "Regulatory Pathways for Clearance or Approvals of IVDs," in FOOD AND DRUG LAW REGULATION 43, 45 (FDLI, eds, 2008). The existing device may include products that were already on the market prior to 1976 and were never subjected to FDA regulation, or those that were cleared after 1976 through a 510(k) process.

existing devices that also were not required to go through the more rigorous PMA process.[6] In contrast to the PMA process, which generally requires submission of high quality clinical data, only 10–15 percent of products cleared through the 510(k) process are required to submit clinical data of any type, making this a much faster and less expensive pathway to market.[7]

For developers of new molecular diagnostics that will be evaluated by the FDA, the agency's decision about which pathway the devices must follow – 510(k) or PMA – has major consequences.[8] While the majority of existing IVD products on the market today have been cleared through the 510(k) pathway rather than through a PMA,[9] the novelty of advanced molecular diagnostics increases the chances that they will be sent through the PMA process as there are less likely to be existing devices to serve as comparators within the 510(k) clearance process, at least initially. Given the substantial time and cost differences inherent to the two different pathways, the lack of predictability creates major problems for molecular diagnostic developers, who are unable to plan their long-term budgets, trials, and other logistics involved with bringing their products to market.

In addition to these challenges, the evidentiary requirements for devices to achieve clearance through the 510(k) process are less than clear. Currently, product sponsors rely on related 510(k) submissions and previously published guidance documents to inform their applications. However, there is a common perception that the FDA's 510(k) require-ments are growing more stringent, especially for molecular diagnostic tests, thereby casting doubt on the reliability of these sources. Specific-ally, 510(k) submissions for molecular diagnostic tests are more likely to require the submission of clinical data and face greater statistical scrutiny than submission for other types of devices.[10] Whether required in the

---

[6]   21 U.S.C. § 360c(2)(C)(ii)(II).

[7]   Government Accountability Office, "Medical Devices: Shortcomings in FDA's Premarket Review, Postmarket Surveillance, and Inspections of Device Manufacturing Establishments," GAO-09-370T, June 18, 2009.

[8]   Institute of Medicine, SYSTEMS FOR RESEARCH AND EVALUATION FOR TRANSLATING GENOME-BASED DISCOVERIES FOR HEALTH 49 (2009).

[9]   In its most recently published data (from 2005), the Office of in vitro Diagnostics (OIVD) reported clearing 434 IVDs through a 510(k) while approv-ing just 9 PMAs. FDA OIVD, FY 2005 OIVD Annual Report 23–24 (2005), available at http://www.fda.gov/AboutFDA/CentersOffices/OfficeofMedical ProductsandTobacco/CDRH/CDRHReports/ucm126863.htm (accessed July 19, 2013).

[10]   Gibbs, supra note 5, at 48.

context of a 510(k) submission or a PMA application, there is a lack of consensus about how to conduct studies that will demonstrate the clinical utility of diagnostic tests.[11] In contrast to clinical trials for therapeutics, in which the effect of a new product can be directly measured by patient outcomes, trials for diagnostics are less straightforward. For example, a diagnostic test that helps physicians tailor the dose of a given drug may be only one factor affecting the outcome of a patient taking that drug, which may also be influenced by other medication use, physician interpretation of the test, patient adherence, and other factors. While it is possible to design large trials to account for these variables, such trials would have to be on the scale of those conducted for pharmaceuticals and are likely prohibitively costly for most device companies.[12]

Another major uncertainty for the developers of molecular diagnostics is potential changes in the FDA's evaluation of IVDs. While the FDA claims to have the authority to regulate all types of IVDs, it has historically distinguished between those that are developed and offered within a single laboratory – referred to as "home brew" or laboratory developed tests (LDTs) – and those that are commercially marketed to multiple laboratories – referred to as IVD "test kits." For the first two decades of the FDA's oversight of medical devices, the agency chose not to evaluate LDTs at all, while requiring all test kits to go through the usual device regulation pathways. This choice was based on the fact that tests offered by single laboratories were historically less complex and lower risk than test kits that were widely marketed.[13]

LDTs have traditionally only been required to comply with requirements contained within the Clinical Laboratory Improvement Amendments of 1988 (CLIA), which are enforced by CMS.[14] These requirements include such things as laboratory certification, proficiency

---

[11]   IOM, supra note 8, at 14.

[12]   David Parker, THE ADVERSE IMPACT OF THE US REIMBURSEMENT SYSTEM ON THE DEVELOPMENT AND ADOPTION OF PERSONALIZED MEDICINE DIAGNOSTICS 18 (Personalized Medicine Coalition, 2010).

[13]   FDA/CDRH Public Meeting: Oversight of Laboratory Developed Tests (LDTs), BACKGROUND (July 19–20, 2010), available at: http://www.fda.gov/MedicalDevices/NewsEvents/WorkshopsConferences/ucm212830.htm (accessed July 19, 2013).

[14]   The Clinical Laboratory Improvement Amendments of 1988, 100 Pub. L. No. 578, 102 Stat. 2903 (codified as amended at Certification of Laboratories, 42 U.S.C. § 263a (1994)).

testing, quality control measures, and laboratory inspections.[15] CLIA regulations were written prior to the widespread use of molecular diagnostic testing, so their quality control measures, proficiency testing, and other requirements were not designed for the complex diagnostic tests being used today. The vast majority of genetic tests marketed today are approved through the CLIA LDT pathway rather than the FDA IVD route.[16] In the past few years, the FDA has started to shift towards regulating LDTs, recognizing a need for providing more oversight for the advanced molecular diagnostics that are increasingly escaping FDA oversight through the LDT pathway. While the agency has not provided a specific timeline for implementing its new regulatory approach, the agency's slow expansion of LDT oversight suggest that it is a question of when, not if, LDTs will be subject to FDA approval requirements.

The possibility that molecular diagnostic tests could face greater oversight in the future has prompted many to claim that such a change would quash future development of the molecular diagnostics field. As discussed above, the cost of bringing a test to market through the FDA is significantly greater than the cost of bringing it to market subject only to CLIA oversight. Because the reimbursement for molecular diagnostic tests is so poor (discussed below), the added costs of FDA regulation are difficult to recoup, even assuming that the tests work well. The lower likelihood of profitability, many argue, would make investors less likely to invest in these technologies.

10.2.2   Reimbursement for Molecular Diagnostic Tests

Just as is occurring with FDA regulation, the traditional reimbursement schemes for laboratory tests are struggling to keep pace with innovation in the molecular diagnostics field. While private payers have the ability to set their own payments for new tests, most payers follow the decisions made by CMS, making the agency's policies especially influential in the success of these tests.[17] Unfortunately, the lack of flexibility in CMS' current reimbursement system has created major obstacles to the clinical

---

[15]   Centers for Disease Control and Prevention, "Good Laboratory Practices for Molecular Genetic Testing for Heritable Diseases and Conditions," 58 No. RR-6 MORBID. MORTAL. WEEKLY REP. 1, 7–25 (2009).

[16]   Gail H. Javitt, "In Search of a Coherent Framework: Options for FDA Oversight of Genetic Tests," 62 FOOD DRUG L.J. 617, 639 (2007).

[17]   Parker, supra note 12, at 2.

adoption and commercial viability of molecular diagnostics. CMS influences the development and adoption of new technologies through three separate activities – coverage, coding, and payment.

First, in terms of coverage, CMS or its local Medicare administrative contractors (MACs) must make a coverage determination about the new product, in which they review the product and decide whether Medicare will pay for its use when furnished to Medicare beneficiaries. CMS makes just 15–20 national coverage determinations (NCDs) each year through a statutorily prescribed 6- to 9-month process requiring systematic evidence evaluation, publication of a proposed decision, response to public comments, and issuance of a final decision. Most commonly, CMS issues an NCD that provides coverage only in limited circumstances, as supported by the available evidence. In some cases, the agency may also decide to either cover the item or service in all cases, to deny coverage in all cases, or to formally leave coverage at the discretion of local contractors.

By statute, the program is prohibited from covering items and services that do not fall within at least one statutorily defined benefit category and those that are not "reasonable and necessary" for the diagnosis and treatment of Medicare beneficiaries.[18] Notably, these criteria differ from FDA's "safe and effective" criteria for product approval, creating a significant disconnect between the evidentiary requirements of the two agencies. Though the phrase "reasonable and necessary" remains undefined for CMS and local contractors, it has been informally interpreted to require a showing that an item or service improves health outcomes in Medicare beneficiaries.[19] For molecular diagnostic tests, payers such as CMS have increasingly called for demonstrations of clinical utility to make this showing.[20] Similar to the FDA, CMS has provided little specific guidance about how to demonstrate clinical utility for molecular diagnostics. In the absence of such guidance, test developers must either meet with CMS and local contractors individually or use CMS' past reviews of molecular diagnostic tests to generate a rough prediction about what types of evidence will be required for their tests. This results in much variability in the quality of evidence provided by test developers

---

[18]   Social Security Act § 1862(a)(1)(A).

[19]   Peter J. Neumann & Sean R. Tunis, "The Growing Demand for Relevant Outcomes," 362 NEW ENG. J. MED. 377, 377 (2010).

[20]   William J. Canestaro, Lori A Martell, E Robert Wassman and Rick Schatzberg, "Healthcare Payers: A Gate or Translational Bridge to Personalized Medicine?," 9 PERSONALIZED MED. 73, 80 (2012).

and inconsistent coverage decisions that delay patient access to important new technologies.[21]

A major concern voiced by the molecular diagnostics industry is that payers are requiring unrealistic levels of clinical data about new tests but have not provided the necessary guidance for test developers to understand how to generate those data, or how they differ from the data provided to the FDA.[22] In the past, laboratory tests were almost uniformly inexpensive and were therefore reimbursed with little scrutiny from payers, even when they lacked strong supportive data. As a result, there is little precedent for test developers to follow when designing studies to demonstrate the value of their new, higher-cost diagnostics. From the payers' perspective, the fact that there is often no way to clinically verify the accuracy of a molecular diagnostic test creates the need for more rigorous validation of these tests compared to, for example, a diagnostic test of hemoglobin for anemia, which can be verified by checking various clinical signs in the patient.[23] The potential consequences of unreliable test results suggests the need for trials that are more similar to those required for pharmaceuticals than the data-sparse 510(k) applications typical of medical devices.

The greater need for data on clinical utility for coverage decisions on molecular diagnostics ties in with the coding and payment determinations for such products,[24] which are determined independent of the coverage decision. In the coding process, a new product is assigned an identifying code, usually one that is shared by other products that are similar in terms of both their clinical function and the resources required to use them. Within the payment process, the agency sets the amount it will reimburse hospitals or providers who use the product in clinical care, generally paying the same amount for all products within a given code. The traditional method in which these reimbursement decisions are made has hindered the efforts of diagnostic test developers to secure product codes and payment amounts that reflect the value of their products – all of which have impeded progress in the development of those technologies and personalized medicine in general.

---

[21]   Steven Olson and Adam C. Berger, "Genome-Based Diagnostics: Clarifying Pathways to Clinical Use: Workshop Report," in INSTITUTE OF MEDICINE, ROUNDTABLE ON TRANSLATING GENOMIC-BASED RESEARCH FOR HEALTH 1, 1 (2012).

[22]   See, for example, Parker, supra note 12, at 9–10.

[23]   Institute of Medicine, THE VALUE OF GENETIC AND GENOMIC TECHNOLOGIES: A WORKSHOP SUMMARY 24–25 (2010).

[24]   Parker, supra note 12, at 18.

The vast majority of molecular diagnostic tests are performed by clinical laboratories and thus are coded within the clinical laboratory fee schedule, which is a list of amounts that Medicare will reimburse for over 1,100 laboratory tests, organized by codes. The fee schedule, which established prospective pricing for clinical laboratory tests, used charge data from laboratories in 1983 to set prices for the hundreds of distinct laboratory services provided across the country. Notably, most if not all laboratory tests in use in 1983 reflected well-known science that had been used for years. As a result, they were priced as commodities with very low marginal costs rather than products of significant research and development as a growing number of molecular diagnostic tests require today. The original legislation apparently did not foresee this evolution and thus contains no mechanism to pay for novel, high-value tests outside of the fee schedule. The inability to provide value-based re-imbursement for laboratory tests creates a significant disincentive for anybody to invest the time and money needed to develop complex molecular diagnostics, no matter how valuable they are for clinical care.

Not only does the fee schedule fail to account for research and development costs, but it also consistently underestimates the bare cost of performing complex tests and therefore provides less reimbursement than necessary for laboratories to simply recoup their costs. For example, Fragile X testing cost $266 to perform in 2004, but was reimbursed at only $62 by Medicare.[25] The discrepancies between fee schedule esti-mated costs and actual costs is at least partially attributable to the fact that the fee schedule has not been regularly updated to account for inflation since 1984.[26] Thus, payment for most diagnostic tests is based on a fee schedule that is widely perceived to under-reimburse for the costs and value of the test. Funders of personalized medicine research point to these and other uncertainties and problems in the reimbursement system as the key reason for declining investment in the diagnostics industry in recent years.[27]

---

[25] Secretary's Advisory Committee on Genetics, Health, & Society (SACGHS), DEPARTMENT OF HEALTH AND HUMAN SERVICES, COVERAGE AND REIMBURSEMENT OF GENETIC TESTS AND SERVICES 46 (2006).

[26] Id. at 43.

[27] Olson and Berger, supra note 21, at 12.

10.3 TWO INNOVATIVE GOVERNANCE SOLUTIONS

The significant regulatory and reimbursement challenges facing molecular diagnostics present a serious impediment to realizing the promise of personalized medicine. While molecular diagnostics will play a central and indispensable role in the new paradigm of personalized medicine, the pre-existing regulatory and reimbursement frameworks for diagnostic tests stand as obstacles to the development and implementation of the new generation of sophisticated molecular diagnostics. The existing frameworks are too slow, low-paying and uncertain to provide the assurances needed by the developers of molecular diagnostics. Molecular diagnostics therefore are trapped by path dependency, where the pre-existing regulatory frameworks have embedded economic, scientific, power, and institutional assumptions that were well-adapted to the older technology but not the new technology.[28]

A long-term, comprehensive solution to the regulatory and reimbursement challenges facing molecular diagnostics will likely require major statutory changes. Such a process is likely to take many years, exemplifying the "pacing problem."[29] In the meantime, though, there are some innovative strategies that are being considered and applied to try to address some of the obstacles facing molecular diagnostics, specifically the delay in getting products to the market and the uncertainty about evidentiary requirements for both regulatory approval and reimbursement. Specifically, two innovations are Coverage with Evidence Development (CED) and parallel review. Although neither of these initiatives is limited to molecular diagnostics, they are both particularly salient and potentially beneficial for molecular diagnostics.[30] In addition to the support of the agencies pursuing these initiatives, the two programs were both endorsed recently by, for example, the 2012 Presidential National Bioeconomy Blueprint[31] and a recent Institute of Medicine workshop.[32]

---

[28] See W.B. Walker, "Entrapment in Large Technical Systems: Institutional Commitment and Power Relations," 29 RESEARCH POLICY 833 (2000).

[29] Marchant et al., supra note 1.

[30] Rachel A. Lindor et al., "Molecular Diagnostics: Regulatory and Reimbursement Innovation," 5 Nos. 175–8 SCI. TRANSLATIONAL MED. 19–22 (March 2013).

[31] White House, NATIONAL BIOECONOMY BLUEPRINT (April 2012), available at http://www.whitehouse.gov/sites/default/files/microsites/ostp/national_bioeconomy_blueprint_april_2012.pdf (accessed July 19, 2013).

[32] Institute of Medicine, GENOME-BASED DIAGNOSTICS: CLARIFYING PATHWAYS TO CLINICAL USE; WORKSHOP SUMMARY (2012).

10.3.1   Coverage with Evidence Development

Coverage with evidence development enables CMS to temporarily cover new products that are not yet supported by sufficient evidence to meet CMS' "reasonable and necessary" coverage threshold while additional data are generated to inform CMS' long-term coverage decision.[33] CMS first formally identified CED as an additional option for making coverage decisions in 2005, whereby it agrees to provide temporary payment for promising new technologies while clinical data are generated to better inform the agency's longer-term coverage decision. In order to receive reimbursement for their use of the product, providers would be required to participate in a clinical trial or input specific clinical data into a registry, creating a body of clinical evidence that CMS will eventually use when making its long-term coverage decision. Research performed in the context of CED differs from research performed for most new products in that the cost of the product being evaluated (but not the administrative cost) is paid for by CMS, rather than by the product sponsor or private researchers. This cost-sharing mechanism not only helps product sponsors but also gives providers and patients faster access to promising products while reducing the chances that CMS will end up paying for ineffective products.

CMS actually used a CED-style policy (without calling it that) for the first time in 1995 to study the effect of lung volume reduction surgery in the treatment of emphysema.[34] In response to the growing popularity of the reduction surgery and limited data on its effectiveness, CMS collaborated with the National Heart, Lung, and Blood Institute and the Agency for Healthcare Research and Quality (AHRQ) to run a multi-center randomized controlled trial. CMS limited coverage to patients treated at one of 17 clinical sites following the clinical trial protocol established by the National Institutes of Health. The results of the study, published in 2003, revealed that the procedure benefited only a small subset of patients and actually increased the mortality rate for others. Using these data, CMS issued a national coverage determination restricting coverage only to the patient subset identified by the study that benefited from the

---

[33]   Rachel A. Lindor, "Advancing Evidence-Based Medicine by Expanding Coverage with Evidence Development," 52 JURIMETRICS 209, 226 (2012); Sean R. Tunis and Steven D. Pearson, "Coverage Options for Promising Technologies: Medicare's 'Coverage with Evidence Development,'" 25 HEALTH AFFAIRS 1218, 1218–19 (2006).

[34]   Muriel R. Gillick, "Medicare Coverage for Technological Innovations – Time for New Criteria?," 350 NEW ENGL. J. MED. 2199, 2200 (2004).

procedure. It has been estimated that this trial, which required a one-time $35 million outlay for research costs, saves CMS $150 million annually by preventing it from paying for ineffective treatments.[35]

CMS also used data collected through CED to inform its coverage of PET scans in the management of various cancers.[36] Prior to the 2005 decision to use CED, CMS had a non-coverage policy for the use of PET in managing most forms of cancer but received a number of requests for reconsideration in the context of specific cancers. In response to these requests, CMS commissioned AHRQ to perform a formal technology assessment. The inconclusive results of the technology assessment highlighted the need for additional information and led CMS to propose the use of CED, carried out through a registry. By 2009, enough data had been collected to conclude that PET scans did improve at least the initial treatment strategy for certain cancers, though the data remained inconclusive for the later stages of management. Using these data, CMS revised its policies to provide broad coverage for the initial treatment stages of the cancers being evaluated, while maintaining the CED policy for later stages.[37]

In total, CMS has used its innovative CED mechanism less than 20 times since officially launching the CED program in 2005. To date, data from CED have only been used to revise two national coverage determinations, while the other applications have not yet started or are ongoing, including one molecular diagnostic CED study currently evaluating the efficacy genotype-guided dosing of warfarin.[38] The success of some of the initial efforts demonstrated the potential of CED to promote more rational spending within CMS and spurred interest in its continued use. However, more recent applications have failed to lead to coverage changes and have pushed CMS to reconsider its implementation of CED.[39]

In the fall of 2011, CMS removed the guidance document governing the application of CED, announced its intention to revise the policy to make it more relevant and useful, and solicited public comments through January 2012.[40] Comments were specifically requested regarding: (a) the

---

[35]   Tunis and Pearson, supra note 33, at 1220–1.
[36]   Id. at 1222–3.
[37]   Id; see also Lindor, supra note 33, at 227–8.
[38]   CMS, NATIONAL COVERAGE DETERMINATION (NCD) FOR PHARMACOGENOMIC TESTING FOR WARFARIN RESPONSE (90.1) (April 25, 2010).
[39]   See Lindor, supra note 33, at 234–7.
[40]   CMS, CED Public Solicitation (November 7, 2011).

use of CED outside of the national coverage determination process; (b) the impact of CED on the Medicare program; and (c) suggestion for ways to apply CED to maximize its benefit to Medicare beneficiaries.[41] In November 2012, CMS released a new draft CED guidance and requested additional public comments, with the goal of finalizing the guidance in the near future. The overall goal of this revision is to enable CMS to make better informed coverage decisions, in a way that improves the health of Medicare beneficiaries while also reducing barriers to further innovation in health care. Given the impediments to the development and commercialization of molecular diagnostics, an expanded CED program would have particular salience and potential for molecular diagnostics.[42]

10.3.2   Parallel Review

Both the FDA and CMS play key roles in the approval of new medical technologies. In the case of diagnostic devices, the FDA ensures that new devices provide a "reasonable assurance of safety and effectiveness," while CMS determines whether those products are "reasonable and necessary" for the care of Medicare beneficiaries. Though these statutory mandates appear to overlap, they have been interpreted much differently in practice, with the FDA focusing predominantly on safety and effectiveness in the context of controlled clinical trials and CMS focusing primarily on patient outcomes in the real-world settings of clinical care.

As a result of these discrepancies, the information requirements of FDA and CMS have traditionally represented separate hurdles for the sponsors of new products. Because products must be cleared or approved by the FDA before they are covered by CMS, many product sponsors seek to satisfy the evidentiary needs of the FDA before even considering the needs of CMS. This process often results in a significant delay between FDA approval and CMS coverage while product sponsors are generating the additional data needed by CMS. The frequent reliance of third party payers on CMS coverage decisions means this delay can significantly affect the rate of adoption and overall success of new technologies. Moreover, FDA and CMS have created much uncertainty about the type and strength of clinical evidence needed to meet the agencies' respective expectations for approval and reimbursement as a result of the shifting evidentiary expectations discussed above.

---

[41]   Id.
[42]   Lindor et al., supra note 30.

To help address these bottlenecks, the FDA and CMS have recently undertaken a new pilot project initiative for a process innovation called parallel review.[43] Parallel review enables product developers to meet with both CMS and FDA simultaneously early in a product's review process, with the goal of clarifying the agencies' evidentiary expectations and reducing the inefficiencies that often result from addressing the agencies' data needs separately. Initiated by FDA and CMS in the fall of 2011, the parallel review pilot program provides a mechanism for the two agencies to concurrently evaluate certain medical devices for approval and coverage, respectively. Specifically, the program allows CMS to begin its national coverage determination (NCD) process earlier, while the product is still being evaluated by FDA. This voluntary program does not change the review standards of either agency but rather seeks to reduce the inefficiencies that often arise when product sponsors address and fulfill the FDA's evidentiary needs without simultaneously considering whether the same data collection process could be used to address CMS's needs as well. By bringing both agencies to the table with the product sponsor earlier in a product's development, the parallel review process is designed to highlight the similarities and differences between the agencies' data needs and help sponsors avoid performing duplicative and inefficient studies, thereby reducing the time and costs of bringing new products to patients.[44]

In response to concerns expressed by product developers during the public comment period on the proposed parallel review pilot program, the agencies amended the parallel review process in a number of ways to make it more flexible and attractive to the product developers. For example, it gave manufacturers whose products had been accepted into the parallel review pilot project the chance to opt out at any time before a final determination had been made by CMS on coverage. The agencies also assured product developers that they would maintain the confidentiality of data shared between the two agencies. In response to concerns about possible delays and additional red tape in the process, the agencies committed to review applications for the program within 30 days of submission, after which time products would follow the normal FDA review process and timeframes.

The parallel review pilot program is planned to run for two years, although the agencies reserved the right to shorten or lengthen the

---

[43]   CMS and FDA, Pilot Program for Parallel Review of Medical Products, 76 Fed. Reg. 62808-62810 (October 11, 2011).

[44]   D.A. Messner and S.R. Tunis, "Current and Future State of FDA-CMS Parallel Reviews," 91 CLIN. PHARM. THERAPEUT. 383, 384–5 (2012).

program duration as appropriate. At the end of the two year pilot project (November 2013), the agencies are expected to review the results of the pilot project and consider making the parallel review process permanent, with any modifications suggested by the results of the pilot project. The parallel review program is not specific to molecular diagnostics, and several different products have been accepted into the pilot program and are currently under review. One product currently taking part in the pilot is Exact Science's Cologuard, a molecular diagnostic assay for stool markers of colorectal cancer.[45]

10.4 CONCLUSION

Molecular diagnostics present a classic case of the "pacing problem."[46] These products are a rapidly emerging new technology, differing in key ways (for example, cost, value, evidentiary requirements) from the earlier technology of relatively simple diagnostic tests. Molecular diagnostics are the sophisticated and high-profile quarterbacks directing the play of personalized medicine, not the fungible commodity linemen plodding along doing their relatively straightforward assignment without a lot of fanfare or recognition. Yet, the existing regulatory infrastructure is designed for the simple diagnostics of yesteryear, not today's high-technology tests that require different regulatory and reimbursement models. The result is a governance bottleneck that is too slow, low-paying and uncertain to provide the framework and incentives needed to usher in the new era of personalized medicine based on molecular diagnostics.

The most common approaches to addressing such pacing problems with other technologies include restructuring of statutory or regulatory frameworks, or creating alternative governance mechanisms, usually outside the scope of government, such as a voluntary or partnership program. For molecular diagnostics, neither of those approaches seems feasible, at least in the foreseeable future. While the optimal solutions to the regulatory dilemmas facing molecular diagnostics may involve sweeping statutory or regulatory changes, these are less likely to be successful in the current political and economic climate – and certainly not in the short term. Instead, two innovative new approaches within the

---

45   See http://investor.exactsciences.com/2010AR/fda/index.html (accessed July 19, 2013).

46   Marchant et al., supra note 1.

existing regulatory framework have been created and are being implemented by the relevant government agencies (FDA and CMS). Both parallel review and CED are currently open to agency review and modification at this time, creating a window of opportunity to propose changes that have the potential to strengthen the utility of these policies and promote the development and use of high value molecular diagnostic technologies.

While neither CED nor parallel review will apply to all or even most molecular diagnostics, they can speed, coordinate and clarify the paths to market for selected products. These examples of creative innovations within an existing and outdated regulatory framework demonstrate another strategy for addressing the pacing problem. Both CED and parallel review have faced opposition both within and outside government, so their implementation has not been easy or uncontroversial. Nor is it obvious to identify and design such innovations. But the CED and parallel case studies demonstrate that it may be possible to expedite and better align regulatory frameworks to new technologies with creative innovation and leadership within government. It is important that we learn from, support and incentivize this type of innovative governance.

10.5   ACKNOWLEDGMENT

The research and writing of this chapter was supported by National Science Foundation grant (Award SES-0921806).

11. Network security agreements: communications technology governance by other means

Joshua W. Abbott

11.1 INTRODUCTION

It has been said we live in the "Communications Age"[1] – a portentous assertion when in other named "ages" (from iron and bronze to nuclear), those who controlled or most effectively used the eponymous technology attained economic, military, and political ascendancy. If this, then, is the Communications Age, governments can be expected to seek effective governance of communications technology and infrastructure as a paramount objective. Indeed, media reports have increasingly depicted lawmakers and analysts fretting over such matters as Internet freedom,[2] online privacy,[3] and various "cyber" issues, including cyber-security,[4] cybercrime,[5] and cyberwar.[6]

---

[1] For example, Oversight of the Federal Communications Commission: Hearing Before the H. Subcomm. on Commc'ns, Tech., and the Internet of the H. Comm. on Energy and Commerce, 111th Cong. (2009) (statement of Julius Genachowski, Chairman, Federal Communications Commission).

[2] See, for example, Michael Martinez, "Clinton Calls for Global Recognition of Internet Freedom," CNN (February 15, 2011), available at http://www.cnn.com/2011/POLITICS/02/15/clinton.internet/index.html (accessed July 8, 2013).

[3] See, for example, Jessica Guynn, "Petraeus Case Triggers Concerns about Americans' Online Privacy," L.A. TIMES, (November 14, 2012), available at http://articles.latimes.com/2012/nov/14/business/la-fi-tn-petraeus-online-privacy-20121114 (accessed July 8, 2013).

[4] See, for example, Eric Engleman, "Cybersecurity Bill Killed, Paving Way for Executive Order," BLOOMBERG (November 14, 2012), available at http://www.bloomberg.com/news/2012-11-15/cybersecurity-bill-killed-paving-way-for-executive-order.html (accessed July 8, 2013).

With the rapid changes in information and communication technologies, ensuring the security of communication networks presents a singular challenge. The question is how the government can most effectively accomplish the goal of securing the nation's networks. Which governance model would be best suited to engage network operators in laying the conditions necessary for meeting that goal? Following a traditional form of governance, Congress might delegate overall responsibility for the problem to a single regulatory agency. That agency could then investigate all the relevant information, seek public input, and eventually promulgate generally applicable rules, to which network operators would be legally required to adhere. Such regulatory processes, however, can be exceedingly slow and inflexible, especially for rapidly evolving technologies.

To secure the nation's communications infrastructure, US federal agencies have improvised an alternative approach – one less formal and more adaptive – involving the use of network security agreements. These agreements, contractual in form, are negotiated between an inter-agency group of government officials and network operators to address potential national security, law enforcement, and public safety concerns.[7] Notably, the process for forming these agreements was not formally established by any statute, regulation, or executive order. The agreements nevertheless govern what security measures the network operators must take to be permitted to operate in the United States. This chapter explores the rationale, procedural process, and some of the consequences and implications of using network security agreements as a form of technology governance.

11.2 BACKGROUND: TECHNOLOGY, REGULATION, AND NETWORK SECURITY

Smart phones today are sleek, powerful, and alluring. They facilitate so many aspects of modern daily life that many smart phone users cannot

---

[5] See, for example, Josh Rogin, "NSA Chief: Cybercrime Constitutes the 'Greatest Transfer of Wealth in History,'" FOREIGN POL'Y, July 8, 2012.

[6] See, for example, Elisabeth Bumiller and Thom Shanker, "Panetta Warns of Dire Threat of Cyberattack," N.Y. TIMES, October 12, 2012, at A1.

[7] See, for example, In re Applications Filed by Global Crossing Ltd. and Level 3 Communications, Inc. for Consent to Transfer Control, Memorandum Opinion and Order and Declaratory Ruling, 26 F.C.C.R. 14056, 14080, ¶¶ 62, 63 (2011).

help growing attached to their device.[8] Anyone who has fallen in love with a new phone, however, knows the relationship will not last long. The certainty that a newer, better model will soon appear makes the fleeting infatuation little more than a brief fling. When we yield to temptation and upgrade, switching devices requires little effort beyond transferring data and learning some new features. We take little thought for how the new bells and whistles might implicate the intricate system of laws and regulations that govern the use of communication technologies. As regulators know, however, some new technologies can render not just older products, but entire regulatory regimes obsolete.

To be sure, regulating a high-tech industry is no easy task. Regulators face the fundamental challenge of formulating a governance scheme flexible enough to keep pace with changing technologies, yet sufficiently definite and detailed to be effective.[9] Intricate regulatory frameworks can fall into obsolescence nearly as soon as they are adopted.[10] By the time a notice of proposed rulemaking is issued, notice and comment periods have elapsed, comments and reply comments are reviewed, and updated rules are adopted and published in the Federal Register,[11] the new rules may already have fallen more than a generation behind the technology on store shelves.

Rapid technological advances in heavily regulated industries can thus be a bane to regulators and businesses alike. Regulators face the Sisyphean task of applying outdated rules to new, often unanticipated technologies. Determining how to apply old rules frequently requires reasoning by analogy and making judgment calls, causing uncertainty about how the rules will be applied. Particularly disruptive technologies are those that defy easy classification into fixed categories – categories often based on outdated assumptions about the technology landscape. Companies, meanwhile, must somehow divine how old rules might be applied to new products or services they develop. Perhaps the only

---

[8]   See David Milgrim, SIRI & ME: A MODERN LOVE STORY (2012) ("I love you, Siri." "I'll bet you say that to all your gadgets.").

[9]   See generally THE GROWING GAP BETWEEN EMERGING TECHNOLOGIES AND LEGAL-ETHICAL OVERSIGHT: THE PACING PROBLEM (Gary E. Marchant, Braden R. Allenby and Joseph R. Herkert, eds, 2011) (exploring the challenging in governing rapidly changing technologies).

[10]   See Lyn M. Gaudet and Gary E. Marchant, "Administrative Law Tools for More Adaptive and Responsive Regulation," in THE GROWING GAP BETWEEN EMERGING TECHNOLOGIES AND LEGAL-ETHICAL OVERSIGHT: THE PACING PROBLEM 167, 167 (Gary E. Marchant, Braden R. Allenby and Joseph R. Herkert, eds, 2011).

[11]   See Administrative Procedure Act, 5 U.S.C. §§ 551–9, 701–6 (2012).

material beneficiaries from uncertain applicability of outdated rules are hourly-billing regulatory attorneys.

At least two major problems result from technologies outpacing the regulatory frameworks by which they are purportedly governed. First, efforts to resolve uncertainties and inconsistencies can delay the deployment of new technologies, to the detriment of providers, consumers, and the economy as a whole. Such delays can be caused directly – as when regulatory authorization is required before a product can be marketed – or indirectly, as when skittish investors wait for regulatory clarity before making capital outlays for new technologies. As many commentators have observed, regulatory uncertainty can chill investment and inhibit innovation.[12] Second, failure to respond promptly to changes in technology can erode the overall effectiveness of the regulatory scheme, implicating all the problems the regulations were intended to address in the first place.

Information and communication technologies provide some of the best, or, rather, worst examples of these challenges. Few industries are regulated as closely as telecommunications, while the explosive advancements in these technologies in recent years have been unparalleled. Under such conditions, one could expect to find a slew of outdated rules and hapless regulations pervading these industries. In fact, the Telecommunications Act of 1996[13] may itself be a prime example.[14] Adopted as an amendment to the 1934 Communications Act,[15] the very structure of the 1996 act was built on a now-disappearing landscape in which different technologies and industries once remained largely separate, each governed by a different regime.[16] Hence, Title I of the act governs Telecommunications Services; Title II, Broadcast Services; Title III, Cable Services, and so on. Historically, each of these services had its own infrastructure and group of providers; it was generally clear which providers were subject to which provisions of the act. Now, however,

---

[12] See, for example, Gary E. Marchant and Douglas J. Sylvester, "Transnational Models for Regulation of Nanotechnology," 34 J.L. MED. & ETHICS 714, 716 (2006).

[13] Telecommunications Act of 1996, Pub. L. No. 104-104, 110 Stat. 56 (codified as amended at 47 U.S.C. 151 et seq.).

[14] See, for example, Commissioner Kathleen Q. Abernathy, "The Telecommunications Act of 1996: A Case of Regulatory Obsolescence?," 13 COMMLAW CONSPECTUS 235, 241 (2004–2005).

[15] Communications Act of 1934, 48 Stat. 1064 (codified as amended at 47 U.S.C 151 et seq.).

[16] See Abernathy, supra note 14.

so-called technological "convergence" is erasing many of those distinctions.[17] Telecommunication service providers, cable companies, broadcasters, and even satellite operators increasingly offer identical services and compete for the same customers.[18] Cable providers now offer phone services, while telecommunication providers offer television programming; media broadcasters often partner with both.[19] Each may use a different technology platform, but differences among platforms are becoming less relevant to the services they deliver. Thus, it is becoming less clear which of the different provisions of the act apply to which providers and services.

The Federal Communications Commission (FCC) is similarly structured along decreasingly relevant distinctions. Separate FCC bureaus are responsible for media, wireless telecommunications, and wireline competition.[20] Many commonplace activities, however, can involve all three of these areas. Suppose a person views a news program on a mobile device via wireless connection to a fiber-optic network. Which FCC bureau exercises jurisdiction over that transaction? The fact that each has responsibility for a different aspect of such seamless activities merely highlights how excessive complexity can arise when static institutional structures govern dynamic technologies.

Even within the narrower area of communications network security, close regulation and rapid technological change pose daunting challenges to effective governance. Network security by its nature draws heightened oversight. In a hyper-connected world with few meaningful boundaries, critical communication infrastructures are under constant threat, and networks serve both as tools and as targets for hostile governments, criminals, and terrorists. Governance, therefore, focuses primarily on preventing unauthorized access to, and foreign control of, US communications and infrastructure. Among the legal means devised to that end,

---

[17]   See, for example, Bryan Tramont, "The Digital Migration: The Future Is Upon Us," 12 COMMLAW CONSPECTUS 127, 127–9 (2004).

[18]   Many providers now offer a service they sometimes call "triple play," which, for a monthly subscription fee, provides telephone, Internet, and cable television service together in one package. See, for example, Robert W. Crandall, J. Gregory Sidak and Hal J. Singer, "Does Video Delivered over a Telephone Network Require a Cable Franchise?," 59 FED. COMM. L.J. 251, 285 (2006–2007).

[19]   See Commissioner Kathleen Q. Abernathy, "Overview of the Road to Convergence: New Realities Collide with Old Rules," 12 COMMLAW CONSPECTUS 133, 133 (2004).

[20]   See Bureaus and Offices, FEDERAL COMMUNICATIONS COMMISSION, available at www.fcc.gov/bureaus-offices (accessed July 8, 2013).

The Electronic Communications Privacy Act (ECPA)[21] – comprising the Wiretap Act,[22] the Stored Communications Act,[23] and the Pen Register Statute[24] – aims to safeguard electronic communications. Pursuant to other laws and regulations, the Committee on Foreign Investments in the United States (CFIUS) reviews investments in and acquisitions of US companies by foreign entities for national security concerns.[25]

The primary regulatory authority over communication networks is the FCC. An independent federal agency, the FCC has comprehensive, congressionally-delegated authority to approve applications to provide telecommunication services.[26] The statutory standard by which the FCC evaluates applications is whether the proposed service would be in "the public interest," a standard that has proven to be as elastic as it is indefinite.[27]

The FCC, however, does not always rely solely on its own expertise in determining which applications promote the "public interest." It has long deferred to the Executive Branch for certain applications on issues of national security, law enforcement, and foreign policy concerns.[28] Over a

---

[21] Electronic Communications Privacy Act of 1986, Pub. L. No. 99-508, 100 Stat. 1848 (1986).

[22] 18 U.S.C. §§ 2510–2522.

[23] 18 U.S.C. §§ 2701–2712.

[24] 18 U.S.C. §§ 3121–3127.

[25] "CFIUS operates pursuant to section 721 of the Defense Production Act of 1950, as amended by the Foreign Investment and National Security Act of 2007 (FINSA) (section 721) and as implemented by Executive Order 11858, as amended, and regulations at 31 C.F.R. Part 800." Office of Investment Security, Committee on Foreign Investment in the United States (CIFIUS), US DEPARTMENT OF THE TREASURY (updated November 13, 2010), available at http://www.treasury.gov/about/organizational-structure/offices/International-Affairs/Pages/cfius-index.aspx (accessed July 8, 2013).

[26] See 47 U.S.C. § 214.

[27] See, for example, Sara Jerome, "Commissioner McDowell Discusses FCC's 'Public Interest Standard,'" HILLICON VALLEY (April 8, 2011), available at http://thehill.com/blogs/hillicon-valley/technology/154917-commissioner mcdowell-discusses-the-fcc-public-interest-standard (accessed July 8, 2013).

[28] See In re Rules and Policies on Foreign Participation in the US Telecommunications Mkt., 12 F.C.C.R. 23891, 23919–20 (1997) [hereinafter Foreign Participation Order]; In re Amendment of the Comm'n's Regulatory Policies to Allow Non-US Licensed Space Stations to Provide Domestic & Int'l Satellite Serv. in the United States, 12 F.C.C.R. 24094, 24170–72 (1997); In re Mkt. Entry and Regulation of Foreign-Affiliated Entities, Report and Order, 11 F.C.C.R. 3873, 3955 ¶ 219 (1995) [hereinafter Foreign Carrier Entry Order].

number of years, this deference has evolved into a semi-regularized process of review by an ad-hoc interagency group, known informally as Team Telecom.[29] It bears emphasis that this process of deferral and review is not directed, mandated, or even addressed by any statute. The FCC conferred quasi-official status on the process by stating it as FCC policy in a 1997 order concerning foreign carriers connecting to US networks.[30] Team Telecom's informal yet compulsory review focuses on the physical and electronic security of an applicant's network infrastructure. The review frequently results in the agencies negotiating a network security agreement (NSA) with the applicant. These agreements include commitments by the applicant to maintain various security measures, including ongoing compliance monitoring by Team Telecom. Grant of the application is then conditioned on adherence to NSA commitments.[31] NSAs thus represent an innovative, though not uncontroversial,[32] form of governance for communication networks.

11.3 THE TEAM TELECOM REVIEW PROCESS

To assess the network security review process as a governance model, it first will be necessary to understand how the process generally works. The Communications Act, as amended, requires all would-be telecommunication service providers in the United States to obtain authorization from the FCC.[33] Over time, the FCC has developed various standards and

---

29 See, for example, National Security Division, US DEPARTMENT OF JUSTICE, FY 2011 PERFORMANCE BUDGET CONGRESSIONAL SUBMISSION 41 (2011), available at http://www.justice.gov/jmd/2011justification/office/fy11-nsd-justification.doc (accessed July 8, 2013).

30 Foreign Participation Order, supra note 28, ¶¶ 61–63.

31 See, for example, In re Applications Filed by Global Crossing, Ltd. and Level 3 Communications, Inc. for Consent to Transfer Control, Memorandum Opinion and Order and Declaratory Ruling, 26 F.C.C.R. 14056, 14080, ¶ 63 (2011).

32 See, for example, In re AT&T Corp., British Telecomm., Plc, VLT Co., Violet License Co., & TNV [Bahamas] Ltd. Applications for Grant of Section 214 Authority, 14 F.C.C.R. 19140 (1999) (separate statement of Comm'r Harold Furchtgott-Roth); In re Applications of Voicestream Wireless Corp., for Consent to Transfer of Control and Assignment of Licenses and Authorizations, Memorandum Opinion & Order, 15 F.C.C.R. 3341, 3383–4 (2000) (separate statement of Comm'r Harold Furchtgott-Roth); Bryan N. Tramont, Too Much Power, Too Little Restraint: How the FCC Expands Its Reach Through Unenforceable and Unwieldy "Voluntary" Agreements, 53 FED. COMM. L.J. 49, 63–4 (2000).

33 See 47 U.S.C. § 214; see also 47 C.F.R. § 63.18.

mechanisms for evaluating whether granting any particular application would serve the "public interest."[34] That determination has evolved into a multi-faceted review, examining the likely effects the proposed service or transaction would have on competition, consumers, and other interests.[35] For certain applications that raise potential issues related to foreign ownership or interconnections with foreign networks, the FCC looks to the Executive Branch to address those concerns.[36]

In addition to economic and regulatory factors, the FCC considers national security, law enforcement, and public safety concerns.[37] In doing so, it seeks to avoid policy conflicts by deferring to Executive Branch agencies with respect to these factors for certain applications – specifically, those reviewed under Sections 214 and 310(b)(4) of the Communications Act.[38] Section 214 governs applications for a license to provide telecommunications services. A party seeking to provide telecommunication services between the US and points outside the US must file an "International Section 214" application.[39] Section 310(b)(4) governs applications to transfer control of wireless licenses to entities with more than 25 percent foreign ownership.[40] In other words, the FCC defers on applications involving international services or substantial foreign ownership of a US wireless license holder, often including satellite networks, undersea cables, and internationally affiliated carriers.

When the FCC receives an International Section 214 or Section 310(b)(4) application, it notifies the Executive Branch as represented by Team Telecom, the ad-hoc interagency group representing the Department of Justice, the Department of Defense, the Federal Bureau of Investigation, and the Department of Homeland Security (collectively, the "Agencies").[41] Team Telecom does not operate pursuant to any particular statute or regulation, nor does it promulgate any regulations itself.[42] No official actions are taken by Team Telecom as such, only under the names

---

[34] See Foreign Participation Order, supra note 28, ¶¶ 44–46.
[35] Id. ¶¶ 47–58.
[36] Id. ¶¶ 61–63.
[37] Id. ¶ 59.
[38] Id. ¶¶ 61–63.
[39] See 47 U.S.C. § 214; see also 47 C.F.R. § 63.18.
[40] See 47 U.S.C. § 310(b)(4).
[41] National Security Division, supra note 29, at 41.
[42] Kent Bressie, MORE UNWRITTEN RULES: DEVELOPMENTS IN US NATIONAL SECURITY REGULATION OF UNDERSEA CABLE SYSTEMS 5 (2009), available at http://www.wiltshiregrannis.com/siteFiles/News/7DF1 C8D035660E8FBEF0AAC7BA8DA103.pdf (accessed July 8, 2013).

of its sponsoring agencies.[43] Nevertheless, Team Telecom exercises significant discretion and authority and operates with minimal transparency.[44]

After receiving notice of an application, Team Telecom may send the FCC a letter requesting it to defer action on the application until Team Telecom has completed its review.[45] Obligingly, the FCC suspends its own review process until it hears back from Team Telecom – a delay that can last several months or even years.[46] Although the FCC normally comports with specified timeframes and deadlines when processing applications, nothing prevents it from deferring a decision indefinitely while awaiting the results of Team Telecom's review.[47]

Team Telecom's purpose in reviewing FCC applications – as reflected in the review process itself – is to secure US communications, to safeguard network infrastructure, to ensure US government access for lawful surveillance, and to detect and prevent criminal and terrorist

---

[43]   See, for example, Petition to Adopt Conditions to Authorizations, In re Yipes Enter. Servs., Inc., Application for Consent to Transfer Control of a Company Holding an International Authorization and Blanket Domestic Authorization Pursuant to Section 214 of the Communications Act of 1934, as Amended, WB Docket No. 07-225, File No. ITC-T/C-20071002-0040700407 (filed on Nov. 30, 2007) (signed by Stewart A. Baker under his title as Assistant Secretary for Policy, US Department of Homeland Security, and by Kenneth L. Wainstein under his title as Assistant Attorney General for National Security, US Department of Justice).

[44]   Bressie, supra note 42, at 5.

[45]   See, for example, Letter from Ty Brown, Attorney Advisor, National Security Division, US Department of Justice, to Marlene H. Dortch, Secretary, Federal Communications Commission (November 26, 2012) (requesting the FCC to defer action on the merger application of T-Mobile and MetroPCS until DOJ, DHS, and FBI complete their review of the matter for any national security, law enforcement, and public safety issues), available at http://apps.fcc.gov/ecfs/document/view?id=7022065433 (accessed July 8, 2013).

[46]   See Kent Bressie, CONTINUING CHALLENGES WITH NATIONAL SECURITY REVIEWS, LICENSING, AND ENVIRONMENTAL REGULATION 10-11 (2011), available at http://www.wiltshiregrannis.com/siteFiles/News/007ADBD538CEA53C7B5B0A6EF6C88283.pdf (accessed July 8, 2013).

[47]   Indeed, many observers, including some FCC Commissioners, have expressed frustration at the delays caused by Team Telecom reviews. See, for example, Voicestream Wireless, supra note 32, n. 3 (Furchtgott-Roth Statement) ("It continues to baffle me as to why the FBI and DOJ are incapable of meeting [the FCC's] filing deadlines. See, for example, SatCom Systems (DOJ/FBI Comments filed 11 months late). Such disregard for [the FCC's] rules undermines orderly decision making and inhibits other parties' abilities to respond to the issues raised in these proceedings.")

activities.[48] In pursuit of these aims, Team Telecom examines the application materials and may have the applicant complete a Triage Questionnaire to determine if a more searching review is warranted.[49] Typical Triage Questions explore the applicant's relationships with foreign entities; who is allowed access to its facilities, network, or data; details about its security systems and policies; its practices of compliance with applicable laws and regulations; information about its business operations and finances; the types of services it provides; and detailed information about its network operations and infrastructure.[50]

If Team Telecom determines upon initial review that an application raises national security, law enforcement, or public safety concerns, it may negotiate an NSA with the applicant. In some circumstances, Team Telecom may decide instead to require from the applicant only a "letter of assurances"[51] – a unilateral commitment from the applicant addressing a few discrete issues of concern – or to revise or renegotiate an existing NSA.[52] Full NSA negotiations, on the other hand, can be lengthy, comprehensive, and detailed.[53]

While terms of individual NSAs vary, most share a number of provisions in common, which have evolved over time as technologies change and as the Agencies continue to develop security policies and priorities. NSAs generally focus on two main subjects: information

---

[48]   Bressie, supra note 46, at 4.

[49]   See DOJ Triage Questions: Questions for FCC Applicants Reviewed by Team Telecom (in Declaration of Catherine Wang, app. A), In re Broadview Networks Holdings, Inc., Case No. 12-13581 (Bankr. S.D.N.Y. September 12, 2012), available at http://www.kccllc.net/documents/1213581/12135811209 120000000 00001.pdf (accessed July 8, 2013).

[50]   Id.

[51]   For example, Petition to Adopt Conditions to Authorizations and Licenses, In re Hawaiian Telecom, Inc., FCC File Nos. ITC-T/C-20120716-00183; ISP-PDR-20120716-00003; WC Docket No. 12-206 (filed December 12, 2012).

[52]   For example, Petition to Adopt Conditions to Authorizations and Licenses, In re Chrysaor S.a.r.l., Transferor, and Astrium Holding S.A.S., Transferee, Consolidated Application for Consent to Transfer Control and Petition for Declaratory Ruling, IB Docket No. 11-143 (DA 11-1488), FCC File Nos. ITC-T/C 20110818-00265 et al. (filed December 13, 2011).

[53]   See, for example, Petition to Adopt Conditions on Transfer of Control (and exhibits), In re Robert M. Franklin, Transferor, and Inmarsat Plc, Transferee, Consolidated Application for Consent to Transfer Control of Stratos Global Corp. to Inmarsat Plc, IB Docket No. 08-143, DA 08-1659, FCC File Nos. ITC-T/C-20080618-00276 et al. (filed January 9, 2009), available at http://apps.fcc.gov/ecfs/document/view?id=6520193191 [hereinafter Inmarsat NSA].

security and infrastructure security.[54] Thus, depending on the nature of the network, NSAs typically require applicants to perform some or all of the following: route all communications through secure facilities located in the US; store electronic data in secure US facilities; comply with lawful surveillance requests from US law enforcement agencies; prevent surveillance by any foreign governments or entities; ensure that only authorized US citizens have access to sensitive information and facilities; and comply with ongoing reporting and monitoring requirements.[55]

After concluding an NSA, Team Telecom files a Petition to Adopt Conditions with the FCC under the names of the Agencies.[56] With a copy of the signed agreement attached (and any sensitive or confidential information redacted), the petition states that the Agencies do not object to the FCC granting the application if approval is conditioned on the applicant's ongoing compliance with the NSA.[57] Once approved, the licensee may remain subject indefinitely to reporting requirements, audits, and site inspections by Team Telecom.[58] If the licensee ever returns to the FCC to propose a new service or transaction, Team Telecom must begin the process anew by reviewing the new application and deciding whether to revisit or renegotiate the NSA.[59]

Although NSAs are drafted in the form of "voluntary" agreements, in substance they more closely resemble regulatory mandates. This distinction is highlighted by the contrasts between an NSA negotiation and that of a commercial contract between private entities or even a typical government contract. For instance, private parties to an NSA face an enormous disparity in bargaining position. None chooses voluntarily to enter negotiations with Team Telecom or have the option to walk away if they ever hope to obtain FCC approval for their services.[60] If an applicant finds a proposed provision objectionable, it can "negotiate" more favorable terms only by persuading Team Telecom, through disclosures of highly confidential information, that compliance would be impractical. In exchange for detailed, often costly commitments from applicants, Team Telecom's only substantive promise is not to object to the FCC's grant of the application. Finally, because the FCC conditions grants of licenses on

---

[54]   Bressie, supra note 46, at 5.

[55]   See, for example, Inmarsat NSA, supra note 53.

[56]   Id.

[57]   Id.

[58]   Id. at 13–18, §§ 5.1–5.13 (comprising Article 5 of the Inmarsat NSA, Reporting and Notice).

[59]   Hawaiian Telecom, supra note 51.

[60]   See AT&T Corp., supra note 32.

compliance with the terms of an NSA, those terms become as compulsory as any law, yet highly individualized, as no rule of general applicability ever could be.

11.4 POTENTIAL BENEFITS AND DISADVANTAGES

As the above illustrates, the network security review process exhibits several innovative characteristics. Two components bear especially on technology governance: the use of contractual agreements between a government agency and a private commercial entity as an alternative to formal regulation, and the deference given by one government agency to another to assess concerns within the latter's separate areas of expertise. The first component allows NSAs to be individually tailored for each carrier's particular technology and network architecture. Team Telecom's use of triage questions and a toolbox of remedies – from simple letters of assurances to full NSAs – also allows it to customize the scope of review in each case. In addition, the confidentiality of the negotiations insulates the government from outside pressures compared to more transparent proceedings. The second component – FCC's deferral to the Executive Branch – draws applicants to the negotiating table, incentivizing them to conclude an agreement quickly to move their applications forward. The involvement of multiple agencies in the process can defray the responsibility assumed by any one agency. These dynamics form the basis for assessing the relative merits and weaknesses of the process overall.

Of course, any estimation of pros and cons demands an inquiry of relevant context, perspective, metrics, and so forth. Almost any attribute could be characterized as either a feature or a bug, depending on viewpoint. For example, a scheme that grants sweeping discretionary powers to a regulator might be hailed by consumer advocates while pilloried by business groups, whereas a laissez-faire approach that gives businesses free rein might garner opposite appraisals. One solution to this problem could be to adopt a "net social welfare" approach, in which total gains to a policy's beneficiaries are weighed against total losses of anyone harmed. Many may find such a solution unsatisfying, however, which deems as successful any policy that results in a net societal benefit without regard to fairness in apportioning costs.

Another essential task when judging a policy's merits is to ascertain any plausible alternatives. A regulated entity might protest a regulation as burdensome, only to find a substituting rule to be even more onerous. Thus, the network security review process should be measured against a likely hypothetical, alternative scenario, such as a formal regulatory procedure. Some might reasonably argue that the current process should be compared instead to a regime in which national security and similar concerns play no part in the FCC's public interest analysis. The intent here, however, is to assess the current process as an alternative form of governance, rather than as an alternative to non-governance. The issue of what degree of regulatory intrusion is most appropriate lies outside the scope of this discussion. It will be assumed, therefore, that in the absence of Team Telecom reviews, the FCC would adopt other means of pursuing the same substantive policies.

With those qualifications in mind, the following points outline some potential benefits of the network security review process as a governance method:

- *Adaptable to new and changing technologies.* A principal advantage of this model is Team Telecom's ability to respond in real time to technological advances.[61] Because it examines each applicant's network individually, Team Telecom can account for even small, incremental changes in its negotiations. In contrast, traditional regulatory rulemaking proceedings must anticipate developments well into the future – for as long as it may take for updated rules to be adopted.

- *Customizable for each applicant's network.* Similar to the first point, individually negotiated NSAs can account for each carrier's particular technology platform, network architecture, and other technical idiosyncrasies. Blanket regulations for an entire sector or class of networks offer no such flexibility. By extension, customized NSAs preclude uncertainty over whether a particular rule provision applies to a given carrier.

- *Potentially more consistent outcomes and enforcement.* Although general policy aims, such as protecting national security, do not vary by technology, the means of best achieving those aims can vary dramatically across different technologies. A technical security standard that proves effective for one network may be completely inappropriate for a network built on a different platform. By individually tailoring NSAs, however, Team Telecom may be able to accomplish the same result – an adequately secure network – however great the technical differences between systems. The same

---

[61]   See Foreign Government Ownership of American Telecomm. Companies, Hearing Before the Subcomm. on Telecomm., Trade, and Consumer Protection of the House Comm. on Commerce, 106th Cong. 2d Sess. 31 (2000) (testimony of William E. Kennard, Chairman, Federal Communications Commission).

principle applies to enforcement of compliance with security requirements. Rules phrased with overly-specific terms might be circumvented by tweaking network elements to fall outside those terms; while rules stated in overly-general terms could make it difficult to determine objectively whether compliance standards have been met. Individualized NSAs could eliminate ambiguity in such determinations, making enforcement easier and more objective.

- *Leverages cross-agency expertise.* In deferring to the Executive Branch, the FCC implicitly recognizes that other agencies may be better equipped to address certain policy concerns. Delegating part of its public interest analysis to Team Telecom allows the FCC to conserve is own agency resources while ensuring that applicants address certain concerns directly with the agencies that have primary responsibility over those issues.

- *Avoids inter-agency conflicts.* The FCC expressly stated this as a reason for its decision to defer to the Executive Branch on certain concerns.[62] With so many federal agencies exercising concurrent jurisdiction with other agencies,[63] allocating overlapping issues to a single agency's review enhances government efficiency and avoids the confusion and uncertainty that would ensue from inconsistent decisions by different agencies.

- *Responds promptly to new information and changing conditions.* In crafting NSA provisions, technology is not the only factor that evolves over time. Team Telecom is able to react nimbly to changing circumstances, whether economic, social, or political. Its agents, individually and collectively, also gain knowledge and experience with each review and in monitoring ongoing compliance efforts. In contrast to rulemaking proceedings that can take months or years to promulgate rules based on new information, Team Telecom can apply new knowledge immediately.

- *Resistant to political pressures.* Team Telecom exercises considerable discretion and attracts minimal public attention in several ways. First, NSA negotiations are strictly confidential,[64] which

---

[62] See Foreign Participation Order, supra note 28.

[63] For example, proposed mergers between telecommunication carriers are reviewed by both the FCC and the Department of Justice's Antitrust Division. While not acting pursuant to the statute or applying the same standards, their respective reviews overlap significantly.

[64] See Bressie, supra note 42, at 15.

prevents outside interference. Second, unlike typical agency decisions, NSAs – ostensibly voluntary agreements – are not clearly subject to any standard of review. The potential benefit to this approach is that Team Telecom is remarkably free to fashion agreements however it believes will best protect the nation's communication networks, without interference from political interests that may not always align with that goal.

- *Efficient decision making.* With relatively few parties involved in NSA negotiations, and without the numerous procedural requirements of a formal rulemaking proceeding, Team Telecom is capable of acting swiftly and decisively.

As counterweight to these potential benefits, the NSA process also exhibits important disadvantages compared to more formal procedures. One observer concluded that "the approach has been one of stumbling incrementalism and remarkable delay."[65]

- *Unjustifiably long and costly delays.* When the FCC receives a request to defer action on an application until Team Telecom has had time to complete its review, the FCC suspends its own review process until further notice. Team Telecom's review can delay an application for weeks, months, or even years, while the applicant waits in regulatory limbo with mounting expenses and missed economic opportunities.
- *Inconsistent requirements over time and across parties.* If NSAs contain differing provisions, then parties are held to different requirements and standards of conduct. It seems unlikely, to say the least, that NSAs differ only to the extent necessary to account for technical variations among networks. And if the standard provisions have evolved over time, then parties to earlier NSAs have assumed different commitments than those negotiated more recently. Consistency in enforcement would also be difficult as different NSAs contain varying compliance and monitoring commitments.[66]
- *Burdens of compliance unevenly apportioned.* If parties are held to inconsistent requirements, they also will bear disproportionate costs for complying with those requirements. By negotiating terms on a

---

[65]   Tramont, supra note 32, at 64.

[66]   See, for example, Bressie, supra note 46, at 8 (noting that while Team Telecom "had dropped the expensive annual-audit requirement with respect to some recent systems, that requirement has recently resurfaced").

case-by-case basis, with no guidelines on what constitutes reasonable compliance costs for different carriers, Team Telecom has no way to ensure that burdens are distributed proportionally. In short, the NSA process inherently raises questions of fairness.

- *Inefficient use of public resources.* Another problem with conducting individual negotiations is that the agencies must repeat the process for each carrier, incurring the associated expenses each time. A rule of general application has the advantage of needing to be worked out only once for it to apply to all relevant parties.

- *Unresponsive to changing technologies for any given network.* Though Team Telecom may adapt negotiations to advancing technologies, any particular NSA remains fixed from the moment it is signed. The terms of an NSA thus remain in force as written, regardless of changes in the carrier's network technology over time. While NSAs can be reviewed and renegotiated, such occasions typically arise only when the carrier seeks a new FCC authorization. Network changes that might render an NSA outdated do not, by themselves, trigger additional review.

- *Overreach of government power.* While the Communications Act grants the FCC broad authority under the "public interest" rubric, no statutory provision expressly directs or authorizes it to defer to the Executive Branch. Nor is it clear whether the FCC, acting alone, could impose all the requirements NSAs contain[67] – in part because its decisions are subject to judicial review and must either be justified or risk being overturned as an abuse of discretion. Furthermore, the notion that NSAs are "voluntary" agreements is farcical. The FCC does not grant applications while they remain under Team Telecom review, and no telecommunication services are permitted in the US without FCC authorization. Carriers subjected to these reviews thus have no choice but to accede to Team Telecom's terms.[68]

- *Lack of accountability.* Team Telecom enjoys broad discretion and is not directly accountable to any reviewing authority. This is largely due to the unofficial character of its institutional role. Because Team Telecom does not exist officially as a government entity, and its members act only under the names of their respective agencies, no institutional mechanisms exist for reviewing "Team Telecom" activities as such. Because Team Telecom does not

---

[67] Tramont, supra note 32, at 62–3.
[68] Id.

function pursuant to any statute, it is not subject to any statutory provisions for judicial review.

- *Lack of adequate recourse for resolving disputes.* By conditioning grants of license applications on a licensee's ongoing compliance with an NSA, the FCC in effect converts NSA provisions into license conditions. The FCC thus puts itself into a position of having to interpret or enforce what is essentially a "contract between a licensee and a third party."[69] A court can review an NSA only if a party brings an action, and only as a contract dispute, rather than as a review of an agency decision.[70] If a dispute arises before an NSA is signed, applicants have no way to overcome Team Telecom objections and move their application forward.
- *Forecloses constructive public input.* The NSA process entails no opportunity for public observation or comment. Team Telecom must therefore rely solely on the representations of the parties and its own institutional knowledge to craft the terms it believes are most appropriate. Neither industry groups, other government entities, private organizations, nor the American public may contribute to the process of deciding "what constitutes a reasonable set of technical and business mandates" to maintain national security and personal privacy.[71]
- *Diminished predictability.* Finally, since NSA terms are negotiated case-by-case, prior NSAs may not reveal what Team Telecom will require of future applicants. Lack of predictability can seriously impede a carrier's ability to plan or prepare appropriately for future operations and investments. A carrier could formulate a strategic plan based on certain assumptions about likely regulatory hurdles, only to have Team Telecom suddenly raise the bar.

Though formidable, these criticisms should be considered with two things in mind – the imperative to ensure the security of communication networks, and the tradeoffs inherent in any governance model. Thus, the question for policy makers is not whether the process involves costs, inconsistencies, and delays, but how it performs relative to alternative approaches. If circumstances can be ascertained in which this process

---

69   Tramont, supra note 32, at 63.

70   See, for example, Voicestream Wireless, supra note 32, § 3.3 (allowing for judicial review of disputes under Voicestream Wireless's network security agreement).

71   Id.

could work well, similar models might be usefully applied to other areas of governance.

11.5 APPLICATION TO OTHER INDUSTRIES

The difficulty in keeping pace with technology is, of course, not unique to communication networks. Even in industries not traditionally considered innovative, advances in other fields can begin to seep over and disrupt long-established business models.[72] If the network security review process can help maintain effective governance under rampant technological change, perhaps similar approaches could be constructively applied elsewhere – especially if characteristics can be identified that make an industry well suited to such an approach.

As explained above, the network security review process actually entails two distinct components: individually negotiated voluntary agreements between a government entity and a private commercial entity, and deference given by one government agency to another for specified policy concerns in reviewing license applications. Each of these mechanisms carries its own implications for policy makers, and may be considered separately for different types of industries.

The practice of negotiating individual agreements on numerous technical and commercial issues is likely suitable only for a concentrated, closely-regulated industry. Cost and efficiency are the primary concerns with this tool. Even in an industry with few participants, negotiating individually with each could expend enormous resources both for the parties and the government. Doing so would become unfeasible for less concentrated industries.[73] Also, an industry with more firms would likely have less need for customized agreements because it likely would have less technological differentiation. Fungibility, by definition, calls for less customization.

Assuming a fairly concentrated industry, the next considerations might be the degree of regulatory oversight and the pace of technology. An industry operating under few regulatory prescriptions, or one relatively undisrupted by technological change, may not necessitate the type of

---

[72] Consider, for example, the publishing industry, for which printing technology long remained virtually unchanged, until electronic readers (an offshoot of personal computing) recently began displacing print on a massive scale.

[73] An exception to this principle may exist if groups of firms are organized to represent their collective interests; then separate agreements might be warranted for each group.

granular requirements that individual agreements afford. Thus, the more concentrated, innovative, and regulated an industry, the more likely an NSA-like process could be useful.

The second mechanism – the FCC's practice of deferring to Executive Branch agencies on certain concerns – accomplishes at least two aims. First, it helps avoid conflicting policies among agencies,[74] especially where agencies claim overlapping jurisdictions. Second, deferring to other agencies can help leverage inter-agency expertise.[75] Concurrent jurisdiction is common among federal agencies and can arise between a sector-specific agency, like the FCC with regard to communications, and one that focuses on particular policies, like many of the individual divisions within the Department of Justice. While the former may have overall regulatory responsibility for an industry, it might lack the latter's expertise on particular policy concerns that arise across industries. In that situation, exercising deference can spare the expense and delays of duplicative reviews. For example, a corporate merger can trigger reviews by the industry regulator and by the antitrust agencies. Indeed, concurrent reviews of proposed mergers occur regularly in telecommunications, media, transportation, energy, agriculture, and other industries.[76] Such reviews are also frequently criticized as inefficient and wasteful.[77]

Overlapping jurisdiction may, however, sometimes reflect legislative intent to disperse government power among agencies. When that occurs, the agencies might coordinate by mutual agreement to assign certain matters to one or the other, thus avoiding duplicative reviews.[78] Or they

---

[74]   See Foreign Carrier Entry Order, supra note 28, at 3955, ¶ 219.

[75]   See Foreign Participation Order, supra note 28, ¶ 62.

[76]   See, for example, Transportation, Energy, and Agriculture Section, Antitrust Division, US Department of Justice, available at http://www.justice.gov/atr/about/tea.html (accessed July 8, 2013).

[77]   See, for example, Jonathan E. Nuechterlein and Philip J. Weiser, DIGITAL CROSSROADS: AMERICAN TELECOMMUNICATIONS POLICY IN THE INTERNET AGE 426 (2005) ("It is debatable whether the public interest demands these additional, largely unchecked layers of intervention [from the FCC's independent merger review] beyond the basic inquiries already conducted by the Justice Department or FTC – inquiries that are considered more than adequate for other industries.")

[78]   For example, the US Department of Justice and the Federal Trade Commission both have jurisdiction to review proposed mergers for competitive effects, so they decide between themselves which agency will review which mergers. See Overview of the FTC/DOJ Clearance Agreement, Federal Trade Commission, available at http://www.ftc.gov/opa/2002/04/clearanceoverview.shtm (accessed July 8, 2013).

can decide for both to review all matters independently. Granting deference could undermine a deliberate policy not to allow too much power to concentrate in one branch or agency. Thus, not every instance of concurrent jurisdiction may be suited to the network security review model.

Concurrent jurisdiction can also result from technological advancement. Historical distinctions between industries can blur as technologies converge, and jurisdictional lines between their respective regulators can grow hazy. Some technologies even give rise to entirely new industries that defy classification into historic categories. Then the problem may not be jurisdictional overlap but jurisdictional gaps.

One example of possible convergence is the use of smart grid technologies, in which energy producers and consumers employ sophisticated data communication technologies to optimize their use of electric power grids.[79] Although the Federal Energy Regulatory Commission (FERC) exercises direct regulatory jurisdiction over electric utilities,[80] it may lack technical expertise on certain issues arising out of utilities' use of communication networks. Indeed, in September 2012, the FERC created an office dedicated to mitigate cybersecurity threats to the nation's power grid.[81] One can imagine a scenario in which the FERC might solicit advice from the FCC, Team Telecom, or other agencies on governing the use or security of such networks. If the need for rigorous examinations of smart grid networks does arise, the agencies could examine the NSA process to see which aspects of it might serve as a useful model.

In sum, concurrent agency jurisdiction may be a situation that warrants deference by one agency to another on particular concerns. Deference may be especially appropriate when the overlap arises, not by legislative intent, but by converging technology. As for Team Telecom's practice of negotiating individual voluntary agreements with private entities, that approach may be best suited to highly concentrated, closely regulated industries that exhibit rapid technological change.

---

[79] See, for example, Smart Grid, ENERGY.GOV, available at http://energy.gov/oe/technology-development/smart-grid (accessed July 8, 2013); see also SMARTGRID.GOV, available at http://www.smartgrid.gov (accessed July 8, 2013).

[80] See Federal Energy Regulatory Commission, What FERC Does, available at https://www.ferc.gov/about/ferc-does.asp (accessed July 8, 2013).

[81] New FERC Office to Focus on Cyber Security, FERC (September 20, 2012), available at http://www.ferc.gov/media/news-releases/2012/2012-3/09-20-12.asp (accessed July 8, 2013).

11.6 CONCLUSION

When conventional models of governance fail to respond effectively to changing technologies, regulators must consider exploring new approaches. The network security review process, though it carries definite disadvantages, illustrates one such alternative approach. By deferring to the Executive Branch on certain applications for national security, law enforcement, and public safety concerns, the FCC leverages other agencies' expertise and avoids policy conflicts. Team Telecom's use of network security agreements allows for individualized security standards to be crafted for each network, accounting for new and unique technologies and architectures. These practices may serve as useful models for certain other industries and regulators seeking a governance model with the necessary flexibility and specificity to govern effectively in an environment of rapidly changing technology.

12. Robust offshore risk regulation – an assessment of US, UK and Norwegian approaches

Preben H. Lindøe, Michael Baram and John Paterson

12.1 INTRODUCTION

This chapter is derived from a research project addressing the robustness of offshore risk regulation for oil and gas operations on the Norwegian Continental Shelf (NCS). The project has involved evaluating and comparing three leading regulatory regimes and assessing their influence on the safety performance of offshore operators. The focus of this chapter is a comparative study of the Norwegian regulatory approach with the US and UK regimes, their modes of regulation and enforcement, and changes over time in response to major accidents.

From the very beginning oil and gas operations on the NCS were dominated by US companies who brought their organizations, technology, industrial culture and practices with them.[1] For more than 40 years, major US companies have been part of a continuous development of the NCS. The blowout at BP's Deepwater Horizon well in 2010 with the deaths of 11 workers, injuries to 16 others, and the largest offshore spill in history has made the comparative analysis more urgent and relevant. The US is thus the second regime included in our comparative analysis. For different reasons the UK regime is the third regime for the comparative study. Since oil and gas exploration started in the North Sea in the mid-1960s there has been an ongoing interchange of information, experience and regulatory developments between UK and Norwegian regulators.

---

[1]   H. Ryggvik and M. Smith-Solbakken, BLOD, SVETTE OG OLJE (Vol. 3 1997).

The global oil and gas industry has many common features. Major operators, entrepreneurs and sub-contractors are operating with similar equipment and similar exploration, drilling and production procedures. They are subject to similar industrial standards and best practices created by a network of expert actors and industrial groups, and an international scientific and technical community. In all these aspects the offshore industry in the three countries has similar features. By contrast other factors seem to be quite different. The purpose of this chapter is a comparative study of the three regimes by assessing differences in (1) the influence of major accidents, (2) political-administrative and legal structures and (3) participatory approaches and industrial relations.

In addition to narratives of major accidents our analytical framework is the concept of a risk regulation regime advocated by Hood, Rothstein and Baldwin who distinguish between the context and the content of a regime and provide a dynamic model with three control components: information gathering, standard setting and behavior modification.[2] Specific to the themes of this volume, this comparative analysis of the three regulatory regimes provides insights into the relative strengths and weaknesses of regulatory systems that rely more or less on soft law versus hard (or "command and control") law for oversight of a rapidly evolving technology such as offshore oil drilling.

12.2 METHOD

From 2007 to 2012 a project, entitled "Robust Regulation in the Petroleum Industry," was carried out by a group of researchers from Norway, the UK and the US[3] Two sets of information provide the empirical basis for our analysis. The first consists of a portfolio of research projects on technological change, risk regulation and safety management systems with regard to the offshore oil and gas industry. The development of the Norwegian continental shelf under a tripartite system of regulators, industry and unions and with strong involvement of the scientific community has contributed to this portfolio. A smaller body of knowledge regarding UK and US regulators and companies working under their jurisdiction has been used. In addition rapidly expanding

---

[2]   Christopher Hood, Henry Rothstein and Robert Baldwin, THE GOVERNMENT OF RISK (2006).

[3]   See Dag Tomas Sagen Johannesen, ROBUST REGULATION IN THE PETROLEUM SECTOR, University of Stavanger (November 11, 2010), available at http://seros.uis.no/category.php?categoryID=6832 (accessed July 16, 2013).

sources of information have been developed after the Deepwater Horizon disaster.[4] Second, relevant laws, regulations, regulatory guidelines and industrial norms and standards have been assessed.

12.3 ASSESSING THE THREE REGIMES

The assessment of the regimes follows two perspectives that are useful for comparative purposes. The first considers the legal and administrative framework, and the second focuses on three elements of the regulatory process; gathering information on risk, setting norms and standards, and enforcement strategies.

12.3.1 The US Regime

In the US, the legal framework governing activities on the outer continental shelf (OCS) comprises an uncoordinated collection of numerous laws enacted by Congress over more than 200 years. There are laws which establish jurisdiction over the OCS and federal ownership of its mineral resources, divide authority between the states and federal government over coastal waters and submerged land, and govern harbors, navigation, vessels, pipelines, and fishing. Additional laws protect national security interests, the rights of native American peoples, marine mammals, and endangered species; prevent air and water pollution and disposal of toxic waste; require environmental impact studies; and authorize liability for personal injury, property damage, and harms to natural resources.

---

[4] See Final Report, National Commission on the Deepwater Horizon Oil Spill and Offshore Drilling (January 11, 2011), available at http://www.oilspillcommission.gov/final-report/ (accessed July 19, 2013); NATIONAL ACADEMY OF ENGINEERING, MACONDO WELL DEEPWATER HORIZON BLOWOUT: LESSONS FOR IMPROVING OFFSHORE DRILLING SAFETY WASHINGTON (2011); Deepwater Horizon Study Group, Final Report on the Investigation of the Macondo Well Blowout, Center for Catastrophic Risk Management, University of Berkeley (2011), available at http://ccrm.berkeley.edu/pdfs_papers/bea_pdfs/DHSGFinalReport-March2011-tag.pdf (accessed July 19, 2013); NATIONAL ACADEMY OF ENGINEERING, ANALYSIS OF CAUSES OF DEEPWATER HORIZON EXPLOSION, FIRE, AND OIL SPILL TO IDENTIFY MEASURES TO PREVENT SIMILAR ACCIDENTS IN THE FUTURE (2011).

Within this framework, the law authorizing OCS oil and gas operations is the Outer Continental Shelf Lands Act (OCSLA).[5] This law, frequently amended since its enactment in 1953, directs the federal Department of the Interior (DOI) and its Minerals Management Service (MMS) unit to conduct OCS leasing programs, issue permits to companies for exploration and production, and develop and enforce the regulations needed to ensure that these activities are conducted safely.[6] It also assigns regulation of workplace safety on OCS rigs to the Coast Guard (CG). OCSLA discusses the regulated industry only with regard to its duties to comply with the regulations developed by these agencies and the sanctions they would incur for non-compliance.

Because Congress has not fully integrated OCSLA with the many other laws applicable to the OCS, the legal framework for drilling operations is not coherent or harmonized. In addition, some of the other laws are implemented by regulatory programs and agencies with their own detailed rules, procedures and decision-making criteria, resulting in a multitude of legally-enforceable requirements that apply to offshore activities. As a result, implementation of OCSLA is very complex for regulators and companies. Also, there is virtually no union presence in US OCS operations and OCSLA precludes the national Occupational Safety and Health Administration (OSHA) from regulating workplace hazards on the OCS.

The main US regulators, MMS and the CG, have over many years enacted numerous rules and standards which are prescriptive and detailed rather than performance-based. Many of these regulations were adopted or incorporated by reference from the numerous industrial standards and recommended practices originally developed by the leading industrial group, the American Petroleum Institute (API), and other private standards organizations. Although OCSLA provides broadly-stated mandates for regulation, regulators have implemented the mandates by developing detailed prescriptive rules (or adopting and making mandatory detailed industry voluntary standards), and then carried out highly proceduralized, checklist inspections to determine if companies are not in compliance and should be subjected to enforcement proceedings and sanctions.[7]

---

[5]   42 USC. §§ 4321–70h (2012).

[6]   Id.

[7]   Michael Baram, "The US Regulatory Regime for Preventing Major Accidents in Offshore Operations," in P. LINDOE, M. BARAM, O. RENN, eds, RISK GOVERNANCE OF OFFSHORE OIL AND GAS OPERATIONS (2013, Cambridge University Press).

Thus, inspection is designed to be a policing and sanction-threatening function even though it heightens the adversarial relationship between operators and regulators. Also, US regulatory experience across many agencies indicates that this approach often leads to low rates of compliance, with agencies then making exemptions and doing other rule-bending to relax overly rigid or detailed prescriptive requirements. Nevertheless, MMS and the CG had combined to conduct announced and unannounced on-site safety inspections and reviews of an operator's compliance documentation. The CG summary of what inspectors should look for consists of a lengthy "national checklist" for numerous "potential incidents of non-compliance" (PINCs), which for example include 160 PINCs for a drilling rig, and other "verifications" that detailed technical requirements are being met.[8]

The factors causing this "command and control" approach with its mechanistic form of inspection are numerous. One is mistrust of industry. Another factor is that US administrative law provides a regulated industry or company the right to challenge agency rules in a federal court. In the litigious US context, this right is routinely used by industry and often results in a court decision overturning a regulation on grounds that it is ambiguous or arbitrary, lacks a sufficient factual basis, or exceeds the agency's mandate. Fearing this outcome, agencies act defensively and spend much time to carefully develop overly detailed regulations.

Another factor that may cause the prescriptive approach to OCS regulation and mechanistic inspections may be that the scale and complexity of activities to be regulated and inspected is so great (7000 active leases and operations in the Gulf of Mexico in 2010) and agency resources so limited that regulatory efficiency becomes the main objective. This militates against developing performance-based rules which require more expertise and evaluation of company performance. But the danger is that striving for such efficiency may not be consistent with ensuring safety, especially when new uncertainties about the regulated activity are involved, as was the case with deep water drilling at the BP Deepwater Horizon site.

Following the Deepwater Horizon accident incurred by BP, which caused multiple rig worker deaths and injuries and the worst-case spill, numerous studies and analyses to determine how and why this disaster occurred were carried out by a special Presidential Commission, MMS and the CG, and several other organizations. Many liability claims were made in lawsuits, a special compensation fund paid for by BP was

---

[8] Id.

established, and a temporary moratorium on deep water drilling was imposed by presidential order. In addition, MMS was terminated and its functions transferred by DOI: leasing to the newly created Bureau of Ocean Energy Management, Regulation and Enforcement (BOEMRE) and safety regulation to a new Bureau of Safety and Environmental Enforcement (BSEE).[9] These new units of the DoI have added to the trove of MMS regulations they inherited several more stringent prescriptive rules for deep water drilling, based on the findings of the accident investigations and analyses (for example, regarding environmental reviews, blowout prevention, cementing, emergency response and spill containment). Following these actions, the moratorium was lifted and new leasing programs announced for the Gulf of Mexico and off the north Alaska coast, with Presidential assurances given to the public that future drilling can and will be safely done. Also early Congressional enthusiasm for increasing company liability for spills has waned because of concern that it would drive companies away from the US OCS at a time when there is high demand for greater self-sufficiency in oil and gas.

Thus, the post-Deepwater Horizon regulatory regime remains prescriptive and inspection remains mechanistic. The only departure has been the enactment of a new safety regulation which is directed at company management of offshore drilling operations, the Safety and Environmental Management Systems (SEMS) rule.[10] SEMS requires that management develop internal controls and practices which fulfill several broadly defined safety-related functions, and thereby resembles the Norwegian performance-based approach to safety regulation (discussed below). The expectation is that companies will fulfill the functional requisites by, at a minimum, following industry standards and best practices, as is done in Norway. This rule could bring about a more holistic and effective approach to safety management offshore and to workplace safety on rigs in particular. Its success will depend on the quality of the industrial standards and practices to be relied upon, and whether BSEE and CG employ inspectors with the expertise needed to evaluate the unique circumstances and set of safety issues on each rig and

---

[9]   See THE BUREAU OF OCEAN ENERGY MANAGEMENT, REGULATION AND ENFORCEMENT, available at http://www.boemre.gov/ (accessed February 21, 2013).

[10]   Oil and Gas and Sulphur Operations in the Outer Continental Shelf – Safety and Environmental Management Systems, 75 Fed. Reg. 63610-01 (October 15, 2010).

determine if an operator's efforts to fulfill the safety functions are sufficient and effective.

Finally, an important issue that awaits BOEMRE and BSEE is the need to improve upon the currently limited approach to gaining information about safety performance. Reports by the offshore industry on safety performance are voluntary but not currently required except with regard to lost time incidents and air emissions.[11] As a result, the safety data base is incomplete and unreliable. Consensus formulas for gauging industry's safety and environmental performance offshore have been developed and it would be useful for the new agencies to use these as "OCS Performance Measures" to evaluate many aspects of performance annually, provide regulators and the public with performance information, and enable OCS operators to compare their performance with industry "averages" and set priorities for improvement.

12.3.2   The UK Regime

In the UK the authorities responded to the emerging industrial opportunities in the North Sea by passing the Continental Shelf Act in 1964. With the United Nations Convention on the Continental Shelf of 1958 having conferred "sovereign rights" in the continental shelf on coastal states, the 1964 act provided the foundation for the licensing regime by which the UK has for almost 50 years organized its relations with the commercial entities who have explored for and produced offshore hydrocarbons. At the outset, health and safety were also dealt with under the licensing regime, with licensees briefly ordered to follow written instructions in this regard from the Secretary of State and the latter ultimately very briefly instructing the former to follow an industry code of practice. The sinking of the Sea Gem in 1965 awoke both public and political attention by pointing up the shortcomings of this minimalist approach to health and safety. Not only did the rig not fall into any recognized legal category, but a subsequent inquiry highlighted the practical and legal difficulties of regulating safety under what was essentially a contractual system. This was the backdrop for the ultimate passing of the Mineral Working (Offshore Installations) Act in 1971, which in turn provided the framework for a detailed prescriptive approach to offshore health and

---

[11]   Jon Espen Skogdalen, Ingrid B. Ume and Jan Erik Vannem, "Summary: Looking Back and Forward: Could Safety Indicators Have Given Early Warnings About the Deepwater Horizon Accident?," Deepwater Horizon Study Group, available at http://www.docstoc.com/docs/130508405/Summary_final_report_Skogdalen_Utne_Vinnem_RNNP_safety_indicators (accessed July 16, 2013).

safety regulation which was progressively introduced by secondary legislation over the following nine years. An immediate challenge to this approach came in the form of the government-commissioned report in 1972 into health and safety regulation in the UK generally under the chairmanship of Lord Robens.

In essence, this report raised profound questions into the appropriateness of the traditional approach which the offshore industry was now following (pointing out the difficulty of developing a prescriptive regime from the regulator's perspective as well as the problems it created in terms of persuading those who created and controlled risk that it was actually the government's responsibility) and laid the foundations for a wholly new approach onshore in the shape of the Health and Safety at Work, etc. Act of 1974 together with a new dedicated regulator, the Health and Safety Executive (HSE). The blow-out on the Ekofisk Bravo platform in 1977 in the Norwegian sector raised new concerns regarding the risk of a major oil spill. The Burgoyne Committee was established with the purpose of assessing risk regulation, administration and enforcement of safety regulation in relation to offshore oil and gas exploration, development and production on the UK continental shelf. The committee raised important issues, including: how to design law and regulations that could be understood and internalized by the industry; whether the regulation should be anchored in the Department of Energy (DEn) or the newly established HSE; and what should be done about the involvement of the workforce in health and safety. Despite the fact that these were in essence the issues raised by Lord Robens in 1972, the net effect of the Burgoyne report was "business as usual" – the offshore industry continued with a detailed prescriptive approach, implemented by a dedicated regulator (the Petroleum Engineering Division) located within the Department of Energy.

This approach, according the offshore industry a special status compared to practically every onshore industry, continued until the Piper Alpha disaster in 1988. With the Cullen Inquiry severely criticizing both the industry and the regulator, its 106 recommendations included wholesale reform of the regulatory regime. Out went detailed prescription implemented by a dedicated regulator located within the sponsoring government department which was also responsible for licensing. In came a goal-setting approach to regulation in the hands of the HSE, leaving the detail to individual operators to be worked out in the context of a safety case on the basis of formal safety assessment and quantitative risk assessment together with workforce involvement and backed up by independent verification. The legal outcome was the Offshore Safety Act of 1992 followed by safety case and goal-setting regulations.

Twenty years later, the responsibility for the acceptance and ongoing review of safety cases remains with the HSE, an executive non-departmental public body of the Department for Work and Pensions. The Department of Energy and Climate Change (DECC) Oil and Gas Directorate licenses and regulates UK oil and gas activities, while its Offshore Environment Unit deals with environmental regulation and the approval of Oil Pollution Emergency Plans (OPEPs). The Maritime and Coastguard Agency (MCA), an executive agency of the Department for Transport, is responsible for deploying counter pollution measures during an oil spill.

In the UK duty holders are required to report injuries, diseases and dangerous occurrences under the Reporting of Injuries, Diseases and Dangerous Occurrences Regulations of 1995 (RIDDOR), and HSE provides yearly reports upon statistics about offshore accidents, dangerous occurrences and ill health. In addition, HSE's Offshore Division has so-called "key programs." The first one, KP1, was launched in 2000 with the aim to reduce hydrocarbon releases. Key Program 3 was initiated in 2004 out of concern for the risk of major accidents arising from asset integrity problems and focused on the effective management and maintenance of safety critical elements (SCEs). The program involved targeted inspections of nearly 100 offshore installations, about 40 percent of the total. One of the results from the work of KP3 was the establishment of the Installation Integrity Workgroup set up by the industry itself and involving 30 oil companies, contractor organizations and independent verification bodies. Among other activities, the workgroup has developed new key performance indicators (KPIs).

The HSE requires that all operations are covered by detailed *safety cases* in which potential hazards, their consequences, and the methods of controlling any risks are described and explained. The overall responsibility for safety on an installation falls to the Safety Case Duty Holder who appoints an Offshore Installations Manager (OIM) to discharge this responsibility. In the case of mobile drilling rigs, the duty holder is the drilling contractor (for example, the equivalent of Transocean in the Deepwater Horizon incident).

Before an operator brings a drilling rig into the UK or operates a fixed platform, they have to prepare a safety case for the HSE to approve. The operator, or license holder, is subject to separate and additional verification requirements under the Design and Construction Regulations in the form of well examinations carried out by an independent and competent person. All parties involved have legal duties to cooperate with both the OIM and the well operator when the well is under construction. The Safety Case Duty Holder and the well operator must demonstrate how their safety management systems will operate together, who has primacy in an emergency, and who has overall responsibility. The 2007 asset integrity report and the findings of the UK Parliament's Energy and Climate Change Select Committee suggest that while the safety case approach may remain the best option among the alternatives (and especially compared to detailed prescription), questions remain as to whether it is as yet being *implemented* as well as it might. The UK's goal-setting safety regulations allow a flexible approach in the choice of technology and systems to meet safety standards. The UK's goal-setting safety regime requires a systematic approach to the identification of hazards and through the application of quality engineered solutions and systems ensures that risks are reduced to as low as reasonably practicable (ALARP).[12]

12.3.3 The Norwegian Regime

Major and fatal accidents on the NCS, especially the blow-out at Ekofisk Bravo (1977) and the capsizing of the Alexander Kielland (1980) gave momentum to rethinking and redesign of the regulatory principles in Norway for the safety of offshore oil and gas production. Those events initiated political processes and pressure for developing new regulations in which the newly established Norwegian Petroleum Directorate (NPD) played a major role. In the 1970s, Norway developed very stringent labor legislation, which materialized in the Work Environment Act of 1977. A unionized industry with complete collective bargaining rights and a comprehensive network of safety representatives recruited from the unions became mandatory for the offshore industry.[13]

Since the 1980s Norway has been working at developing a coherent, integrated legal framework for regulating health, safety and environment in the conduct of NCS oil and gas operations. Risk regulation has been developed step by step in the direction of increased use of functional requirements expressed in legislation as legal standards. The regulations focus on promoting self-regulation by operators by directly requiring

---

[12]   See John Paterson, BEHIND THE MASK: REGULATING HEALTH AND SAFETY IN BRITAIN'S OFFSHORE OIL AND GAS INDUSTRY (2000); John Paterson, THE EVOLUTION OF OCCUPATIONAL HEALTH AND SAFETY LAW ON THE UK CONTINENTAL SHELF 43–67 (Vol. 27, 2007); Greg Gordon and John Paterson, OIL AND GAS LAW: CURRENT PRACTICE AND EMERGING TRENDS (2011).

[13]   J. E. Karlsen and P. H. Lindøe, The Nordic OHS Model at a Turning Point?, 4 POLICY AND PRACTICE IN HEALTH AND SAFETY 17, 17–30 (2006).

each operator to develop and apply an "internal control" system for reducing risks and preventing and responding to accidents, a system which reflects "a sound health, environment and safety culture".[14]

Norway, pushed by the Organisation for Economic Co-operation and Development (OECD), made the same efforts as others in modernizing their regulatory and supervisory agencies in general.[15] Even as the need for improved coordination, harmonization and reduction of the number of regulatory agencies has been recognized as important by industries, it has been difficult to agree upon the principles that should be applied in doing so.[16] However, there is one exception in the oil and gas industry. All safety, health and environment control functions founded in different laws and ministries have been delegated to the Petroleum Safety Authority (PSA) as was done to the broader HSE in the UK

Since 1972 the Norwegian Petroleum Directorate (NPD) has been a strong coordinating regulator. In 2004, as part of a comprehensive re-structuring of the regulatory system in Norway, safety regulation was transferred to a new agency, the PSA.[17] This decision by Parliament had two important consequences for the petroleum regime: (1) a transfer of the responsibility for a number of petroleum related land facilities to the NPD; and (2) a transfer of health and safety issues from the NPD to a new regulatory body, the PSA, leaving the resource management administration with the NPD.[18] From initially having supervisory responsibility of 14–15 major companies operating on the NCS, the PSA currently has supervisory responsibility for 110 companies onshore and offshore ranging from "small independents" to "super majors" such as BP and Shell.

In 1985 an external reference group for regulatory development with authorities, employers and unions was established in Norway. At the end of the 1990s, this tripartite system was strengthened and expanded by

---

[14] Jacob Kringen, "Culture and Control. Regulation of Risk in the Norwegian Petroleum Industry," Universitetet I Oslo (2006), available at https://www.duo.uio.no/handle/123456789/17876 (accessed July 16, 2013).

[15] Organisation For Economic Co-operation and Development, REGULATORY REFORM IN NORWAY – MODERNISING REGULATORS AND SUPERVISORY AGENCIES (2003), available at http://www.oecd.org/regreform/2955909.pdf (accessed July 16, 2013).

[16] Jan Hovden, "Public Policy and Administration in a Vulnerable Society: Regulatory Reforms Initiated by a Norwegian Commission," 7 JOURNAL OF RISK RESEARCH 629 (2004).

[17] Id. See also P. H. Lindøe and O. E. Olsen, "Responses to Accidents in Different Industrial Sectors," 49 SAFETY SCIENCE 49, 90–7 (2011).

[18] P. H. Lindøe and O. E. Olsen, CONFLICTING GOALS AND REGULATORS ROLES (Vol. 1, 2007).

new arenas for cooperation.[19] The tripartite group was extended by the Pollution Control Authority and the health authorities and labeled the "Regulatory Forum."[20] In the context of offshore safety, an initiative was taken by the authority to create a "Safety Forum" where the most important actors meet regularly. The industry took initiative to two programs: (1) "Working together for safety," addressing activities with high risk potential and making improvements on installations, industrial standards and procedures; and (2) a training program with the purpose giving offshore workers an introduction to rules, standards and legal practice on NCS.

In the Norwegian regime the industry is required by law to report incidents that could lead to severe accidents and to report occupational accidents. The PSA approach has been to develop a monitoring program covering all risk aspects within the PSA's jurisdiction.[21] The first report was presented by PSA early in 2001 based on data for the period 1996–2000. Since then, annual updates have been performed in cooperation with the industry, unions and with support from researchers. The report uses statistical, engineering and social science methods to provide a broad illustration of risk levels, including risks due to major hazards or to incidents that may represent challenges for emergency preparedness, and risk perception and cultural factors. The statistical approach is based on recording the occurrence of near misses and relevant incidents, performance of barriers, and results from risk assessments. Safety culture, motivation, communication and perceived risk are covered through the use of surveys, interviews, audits, inspection reports, and accident and incident investigations.[22]

Within the offshore risk regulation framework the use of legal standards is widely used. That means standards signifying a very explicit norm, even though linguistically quite broadly formulated and seeking its content outside of the legal writings. Certificates, industrial standards regarded as sound professional practice or the "reasonable man standard"

---

[19]   Moen Anita, Blakstad Helene Cecilie, Forseth Ulla & Rosness Ragnar, DISINTEGRATION AND REVIVAL OF TRIPARTITE COLLABORATION ON HSE IN THE NORWEGIAN PETROLEUM INDUSTRY (2009).

[20]   G. S. Braut and P. H. Lindøe, RISK REGULATION IN THE NORTH SEA: A COMMON LAW PERSPECTIVE ON NORWEGIAN LEGISLATION (2008) (presented at the Working On Safety Conference).

[21]   J. E. Vinnem, "Risk Indicators for Major Hazards on Offshore Installations," 48 SAFETY SCIENCE 770, 770–87 (2010).

[22]   See generally PETROLEUMSTILSYNET, available at www.ptil.no (accessed February 21, 2013).

in a specific field of activity are examples. Legal standards, compared to explicit requirements, make the connection between the regulatory framework and the regulated activities more obvious. Further, they permit more updated regulatory practices than is possible when relying solely upon written statutes with detailed content, thereby facilitating technological development. They also support "multilateral" tripartite regulation involving companies, employees, and the regulating authorities. A possible disadvantage is that legal standards may open up a multitude of possible interpretations if regulatory practice through court decisions, supervisory activities or sector involvement is at a minimum level.

Legal standards are widely used within the offshore regulation framework in Norway. The interpretation and practicing of the legal standards is facilitated by the use of the tripartite arenas presented above and reinforced by regulatory oversight (for example, by the PSA). An effective use of legal standards requires mature actors with a high competence and motivation to keep professionally updated continuously.[23] It also results in company adoption of consensus standards and practices for the industrial sector, many of which embody proceduralization. Thus, use of common law concepts does not mean that detailed proceduralization is absent from the Norwegian system. The tripartite cooperation of authorities, operators and labor unions in problem-solving has created a PSA-managed, non-adversarial approach to building safety systems within each company. The regulatory means of enforcement by the PSA give priority to soft regulation, from non-statutory means as "identical letters of interpretations" to "orders."[24]

12.4   ANALYSIS AND DISCUSSION

The United States, United Kingdom and Norway have taken very different approaches to regulating the safety of offshore exploration for and production of oil and gas resources. History and societal and cultural values influence the governmental authorities, industrial actors and the workforce (unions), which became obvious in assessing the risk regimes developed in the US and the "North Sea". The US regime can still be characterized as "command and control" with a top-down approach where regulators demand that the industry comply with the detailed rules that they set down. By contrast, the UK and Norway follow the principle

---

[23]   See Lindøe, supra note 20.
[24]   Robert Baldwin, Martin Cave and Martin Lodge, UNDERSTANDING REGULATION: THEORY, STRATEGY, AND PRACTICE (2013).

of enforced self-regulation thereby rely on the capability of the industry to manage their own risks according to accepted norms and standards.[25]

The Norwegian regime goes even further in developing a tripartite system based on egalitarian values and mutual trust among involved actors. In its welfare state model, Norway promotes a symmetrical partnership between public agencies and industrial actors which involves labor unions, in parallel with the asymmetric role of sanctioning industry for violations of law. This differs from a command and control regime with regulators requiring that industry must comply with the regulator's rules or be punished. In the process of redesigning the regime after 2010 the US regulators considered the UK and Norway regimes even though there are voices questioning why the UK "safety cases" and Norway's trust-based soft path to industrial compliance should come to America.[26]

A summary of some of the different characteristics of the US, UK and Norwegian regimes is presented in Table 12.1.

The disparate development of offshore regulatory regimes has been influenced in significant part by major accidents. The interconnection and relations between major accidents and the development of the three regimes is presented in Table 12.2. From the Sea Gem accident in 1965 to the Deepwater Horizon incident 45 years later, existing regimes have been challenged and changed by the political and administrative process that followed these significant accidents. This pattern is most obvious in the North Sea. The blowout on the Bravo platform in 1977 initiated the Burgoyne Committee in the UK, while the catastrophe with the Alexander Kielland flotel three years later turned the Norwegian regime away from prescriptive rules inherited from maritime regulation. In turn, the UK regulator could draw on the Norwegian experience in the redesign of their regime in the early 1990s after the Piper Alpha disaster.

On the US OCS, no major drilling accident occurred for 40 years following the 1969 blowout and spill of the Amoco Cadiz until BP's Deepwater Horizon in 2010. This may be one reason that the US regime was not challenged or changed over that period. The numerous lesser incidents and non-major accidents that did occur were tolerated, and

---

[25] A. Hopkins, "Risk-management and Rule Compliance: Decision-making in Hazardous Industries," 49 SAFETY SCIENCE 110, 110–20 (2011).

[26] Rena Steinzor, "Lessons from the North Sea: Should 'Safety Cases' Come to America?," 38 B.C. ENVTL. AFF. L. REV. 417; M. Baram, PREVENTING ACCIDENTS IN OFFSHORE OIL AND GAS OPERATIONS: THE US APPROACH AND SOME CONTRASTING FEATURES OF THE NORWEGIAN APPROACH, University Of Stavanger and Norwegian Research Council (October 2010).

Table 12.1   Characteristics of Control Components in the Three Regimes

Regime	Information gathering	Standard setting	Behavior modification
USOCS	Legal requirement of lost time injury, discharge and emission, but not yearly updating of safety performance data. Initiatives now taken to improve voluntarily reporting.	Laws and regulations with prescriptive detailed rules providing a multitude of legally-enforceable requirements with industrial standards included.	Unannounced and announced inspections using detailed checklists of "potential incidents of non-compliance" (PINC). Hard policing and sanctions for non-compliance. Low involvement of workers and unions.
UKCS	Requirement to report injuries, diseases and dangerous occurrences. Yearly reports and statistics provided by HSE. The "key program" provides important safety indicators.	Goal and risk based regulation with a detailed "safety case" has to be qualified by independent and competent actor and approved by HSE.	A flexible approach balancing enforcement with the industry's choice of technology and systems to meet safety standards.
NCS	A monitoring program of safety performance, based on tripartite effort has been developed since 2001. Gives priority for regulator enforcement strategy.	Coherent and integrated laws and regulations. Risk and performance-based with use of legal standards with flexible interpretation and use of industrial standards.	Based on dialogue, trust-based and soft instruments as enforcement strategy. Involvement of workforce unions at national, industrial and company level.

when they caused demands for protection by coastal states, presidential and other moratoria were decreed in the relevant regions and these satisfied the critics. Thus the need for regulatory reform was not felt at the federal level until Deepwater Horizon, a worst-case major accident which provoked international outrage. Since Deepwater Horizon, the US has been considering and making several changes in its regulatory regime, and the EU and offshore oil producing countries are also re-evaluating their own regimes. Many are looking into the UK and Norwegian regimes to see if there are lessons to be learnt in building better offshore safety regimes.

Table 12.2   Major Accidents and Phases of Regulatory Regimes

Time	Major accident	UK regulations	Norwegian regulations	US regulations
1961–1970	Sea Gem (1965), Amoco Cadiz (1969)	Continental Shelf Act (1964)	Petroleum Act (1963)	Outer Continental Shelf Lands Act (1953)
1971–1980	Bravo (1977) Alexander Kielland (1980)	Mineral Working (Offshore Installation) Act (1971), Robens report (1972), HSWA (1974), Burgoyne Committee (1977)	Regulations relating to safe practices (1975 and 1976). Work Environment Act (1977)	
1981–1990	Piper Alpha (1988)	The Lord Cullen Report, 1990	Principles of internal control (1981), Petroleum Act (1985)	
1991–2000		Offshore Safety Act (1992)	Petroleum Act (1996)	
2001–2011	BP Deepwater Horizon (2010)	Offshore Installation (Safety Case) Regulations (2005)	Revised regulations (2011)	Separation of leasing function and replacement of MMS with BOEMRE and BSEE, same but some new prescriptive rules and SEMS rule (2010, 2011)

A particularly interesting aspect is the dynamics between onshore and offshore regulation in the UK compared to Norway. As mentioned above, the Robens Committee on Health and Safety at Work (1972) came up with recommendation which were in stark contrast to the regime which had just been set up for the UK offshore industry under the Mineral Workings (Offshore Installations) Act of 1971. Lord Robens concluded that there existed too much detailed prescription of every aspect of work as purely a matter of government regulation and not of individual responsibility; too much of the existing law was irrelevant to real problems; and there was a major disadvantage in attempting to address the problems with the wide array of administrative agencies then engaged in the field. This recommendation became embedded in the new Health and Safety at Work Act from 1974 and led to the establishment of the HSE. It nevertheless took fully 20 years for these recommendations to be implemented fully on the UKCS in the aftermath of the Piper Alpha disaster. By contrast, the Robens report and subsequent Health and Safety at Work Act influenced the Norwegian regime both onshore and offshore in the direction of enforced self-regulation or internal control.[27]

12.5   CONCLUSION

The purpose of this chapter has been to compare differences between the US, UK and Norwegian offshore regulatory regimes by assessing how major accidents affect the regulation, political-administrative and legal structures and participatory approaches among the actors including industrial relations. The lesson learnt from the study can be summed up in some specific findings regarding the three regimes.

The first finding is related to how major accidents influenced the regulations. When the emergent industrial development of oil and gas in the North Sea took place from the 1960s an effective regulatory framework to control and enforce safety within the industry wasn't in place. From the sinking of the Sea Gem platform in 1965 to the Piper Alpha disaster in 1988 major accidents became eye openers for politicians and the public as well as triggering agents in the development of new and better rules and regulations in the UK and Norway. Contextual factors were common in the two jurisdictions: the hazardous marine environment, plus close distance between the fields with exchange of mobile platforms and vessels. The relative short time lag between the accidents created a dynamic in interchange of ideas and experience between the regulators of the two nations. This picture is very different from the US where no major drilling accident occurred for 40 years between 1969 until BP's Deepwater Horizon in 2010. The US regime seems not to have been really challenged over that period.

The second finding is related to the political-administrative and legal structures. From the 1970s the political process and legal adjustment in

---

[27]   See Karlsen, supra note 13.

UK was initiated by the accidents and included the administrative responsibility of regulation of resources and safety. An agency responsible for offshore oil and gas resources in Norway was created in 1972, including safety, and the arrangement was not changed until 2004. In the US the split of offshore resource and safety regulation came after Deepwater Horizon in 2010. In the UK and Norway new principles regarding occupational health and safety onshore in the direction of enforced self-regulation or internal control were transferred to offshore regulation. In Norway the process took place in the 1980s while it took 20 years until the recommendations were implemented on the UKCS in the aftermath of the Piper Alpha disaster. Opposite to the UK and Norway, the principles of "command and control" have continued in effect in the US, with the 2010 Deepwater Horizon accident triggering the SEMS rule and its focus on internal control by industry.

Third, it becomes obvious that the tripartite system of cooperation and balancing of interests among the authority, industry and workforce/unions developed in Norway and to some extent in the UK differs from the system in US One core element in the tripartite system is the trust developed between the regulator and the regulated industry and their recognition of the legitimate role of the workforce as a basis for industrial relations. That enables the regulator to implement a strategy in which some of the major responsibilities for control are shifted to the industry and there is a presumption that there is willingness and capability to collaborate upon a continuous development of accepted norms and standards.

The transfer of operators, management and technology between globally operating firms may imply the risk of new failures, misuse and accidents.[28] A systematic assessment of the societal, economic and industrial framing for such transfer seems to be needed in order to determine if a particular regulatory approach is optimal in a specific context. A strategy of enforced self-regulation with a "bottom-up" strategy of developing industrial standards may be effective where it is likely that consensus can be achieved among actors with long lasting relations. In contrast, if there is a continuing need or desire to provide stable norms and controls in an unambiguous way by regulators, a "top-down" strategy with prescriptive rules and regulations and sanctions would be better. There is need for more empirical study and reflection regarding different regulatory regimes and their interplay and effects on

---

28   O. E. Olsen and P. H. Lindøe, "Risk on the Ramble: The International Transfer of Risk and Vulnerability," 47 SAFETY SCIENCE 743, 743–55.

safety management systems as well as the legitimate role of the stake-holder.[29] Such study and reflection may improve the development of robust risk regulation and governance applicable to a wider range of industrial sectors.

12.6 ACKNOWLEDGMENTS

The chapter is based on research from the Petromaks program funded by The Research Council of Norway http://www.forskningsradet.no/prognett-petromaks/Nyheter/The_PETROMAKS_programme_A_major_boost_for_petroleum_research/1253981434837/p1226993690988 (accessed July 19, 2013).

---

[29] See Responses to Accidents, supra note 17.

13. Conclusion: emerging governance for emergent technologies

Gary E. Marchant

This book has addressed the governance of emerging technologies. It has examined a number of different technologies, from railroads to robotics, biotechnology to synthetic biology, molecular diagnostics to nanotechnology, and network security to offshore oil well safety. It has looked at different jurisdictions – including the United States, Norway, the United Kingdom, Canada and Australia. The one clear, consistent lesson that emerges from considering the governance of these various technologies in numerous jurisdictions is that traditional government regulation is woefully and inevitably inadequate for managing the risks of emerging technologies.

Emerging technologies present several challenges to traditional regulation. First, many emerging technologies, such as nanotechnology, biotechnology, and robotics are not limited to a single industrial sector like were many previous technologies (for example, motor vehicles, pharmaceuticals, chlorofluorocarbons), but rather span multiple industries and a multitude of applications. This resulting diversity of potentially regulated entities and application contexts complicates regulatory efforts by, for example, making it harder to identify and evaluate all companies involved in the technology life-cycle, and to assess the exposures, risks, and control costs of the many different applications.[1]

Second, many emerging technologies present unprecedented uncertainty with respect to their risks, benefits and future development. For example, traditional toxicity assays may not work for nanomaterials that may cause harm through novel mechanisms and exposure scenarios.[2] To

---

[1] For example, Rebecca Kessler, "Engineered Nanoparticles in Consumer Products: Understanding a New Ingredient," 119 ENVTL. HEALTH PERSP. A121, A122 (2011).

[2] Gary E. Marchant et al., "The Biggest Issues for the Smallest Stuff: Nanotechnology Regulation and Risk Management," 52 JURIMETRICS 243,

some, genetically modified foods present unprecedented and unacceptable risks to public health and the environment, whereas to others this
represents perhaps the safest technology ever created by humans.[3] The
future applications and trajectories of many technologies are highly
uncertain and speculative – for example, synthetic biology might represent a small, incremental enhancement to existing biotechnologies, or it
could result in fundamental transformations of humans and other species.
Information and communication technologies may continue to enhance
the convenience and efficiency of modern lifestyles, or they may lead to
a *Brave New World* in which privacy and freedom are significantly
restricted if not eliminated. As Brad Allenby notes in Chapter 2 with the
historical example of railroads, the long-term consequences, second-,
third- and fourth-order, of deploying a new technology often cannot be
predicted or anticipated.

Third, many emerging technologies raise potential risks that are much
broader than many previous technologies, which generally have been
limited to environmental and health risks. While most emerging technologies also present such safety-related risks, they also create other
types of risks such as privacy risks, human enhancement issues, and
socio-economic concerns. These types of risks are more difficult to
regulate, and often are outside the jurisdiction of existing regulatory
agencies and statutes.[4]

A fourth complication of many emerging technologies is the international dimensions of the technology. Many emerging technologies are
being developed simultaneously in many different industrial nations,
often by the same multinational entities operating in multiple jurisdictions. The increasing global nature of trade and commerce make
technologies international even if they are initially only produced domes-
tically. For these reasons, there is a growing realization and focus on
international coordination and harmonization of technology governance,

---

248–50 (2012); Stephan T. Stern and Scott E. McNeil, "Nanotechnology Safety
Concerns Revisited," 101 TOXICOLOGICAL SCI. 4, 16 (2008).

  [3] Nina V. Fedoroff and Nancy Marie Brown, MENDEL IN THE KITCHEN:
A SCIENTIST'S VIEW OF GENETICALLY MODIFIED FOOD (2004).

  [4] Gary Marchant, Ann Meyer and Megan Scanlon, "Integrating Social and
Ethical Concerns Into Regulatory Decision-Making for Emerging Technologies,"
11 MINNESOTA J. LAW SCIENCE & TECHNOLOGY 345 (2010).

a task for which traditional "hard law" regulation is particularly under-developed and inept.[5] When national regulations of new technologies are not coordinated, the result can be international trade wars, animosity and inefficiencies, as has been experienced with genetically-modified foods.[6]

Fifth, and probably most importantly, most emerging technologies are developing at an exponential rate, quickly leaving behind any traditional regulatory frameworks. This so-called "pacing problem,"[7] in which regulation cannot keep pace with the technology it seeks to regulate, creates two related problems. On one hand, there are often no regulations promulgated to govern rapidly emerging technologies, while on the other hand, any regulations that are promulgated are quickly outdated by the fast-moving technology. Thus, regulators are damned if they do and damned if they don't – if they wait for more certainty, they risk sitting on their hands while watching a new technology potentially cause serious harms, whereas if they attempt to act proactively and adopt regulations, they risk the regulations becoming obsolete and doing more harm than good, perhaps even before they have finished being promulgated.

For all these reasons, traditional government regulation alone cannot adequately govern most emerging technologies. This is the first clear conclusion that all the chapters and authors of this book appear to agree on. While there will still be a role for traditional government regulation, something more is needed. But what can best fill the "governance gap" that afflicts so many emerging technologies? The primary objective of this book is to start to answer that question. Another clear finding is that there is no single approach, process or institution that will work for all technologies. There is no magic bullet. The most appropriate strategy for any given technology in a particular jurisdiction will be highly context-specific, depending on the nature and characteristics of the technology, the pre-existing legal framework, and the social and political environment in which the technology exists. As Preben H. Lindøe, Michael Baram and John Paterson demonstrate in their comparative analysis of the govern-ance of offshore oil drilling in the United States, United Kingdom and

---

[5]   Gary E. Marchant and Kenneth W. Abbott, "International Harmonization of Nanotechnology Governance through 'Soft Law' Approaches," NANOTECH. LAW & BUS. (in press, 2013).

[6]   Mark A. Pollock and Gregory C. Shaffer, WHEN COOPERATION FAILS: THE INTERNATIONAL LAW AND POLITICS OF GENETICALLY MODI-FIED FOODS (2009).

[7]   THE GROWING GAP BETWEEN EMERGING TECHNOLOGIES AND LEGAL-ETHICAL OVERSIGHT: THE PACING PROBLEM (Gary E. March-ant, Braden R. Allenby and Joseph R. Herkert, eds, 2011).

Norway, there are strong drivers of a more flexible, cooperative and adaptive approach than offered by traditional "hard" rules, but the acceptance and feasibility of a more "soft law" approach is dependent on political, legal and cultural milieu in a given jurisdiction. Even within a given jurisdiction, the relative priority given to hard, proactive regulation versus a more lenient and laissez-faire approach might change over time and with different technologies, as Diana M. Bowman shows with her contrast of the Australian government's approach to biotechnology versus nanotechnology.

One important first step to address this complexity is for those involved in the governance of a new technology – including government regulators, industry and non-governmental organizations – to be more anticipatory about the impacts and challenges of an emerging technology. As the pace of new technologies continues to accelerate, passivity and waiting are fatal to effective governance. Anticipatory technology assessment can help to avoid or minimize many of the mistakes of the past, as Timothy F. Malloy argues in his analysis of examples such as the ill-fated regulatory requirement for MTBE in gasoline. The anticipatory technology assessment advocated by Malloy and others can help regulators and stakeholders faced with outmoded regulatory frameworks to develop proactive approaches for overcoming the outdated laws. Two case studies of such innovative approaches described in this book are the efforts of the Food and Drug Administration (FDA) and the Centers for Medicare and Medicaid Services (CMS) to experiment with new regulatory approval and reimbursement innovations for sophisticated new molecular diagnostics, as described by Rachel A. Lindor and Gary E. Marchant, and Joshua W. Abbott's description of how the Executive Branch and the Federal Communications Commission (FCC) are cooperating to address the national security issues raised by telecommunications applicants through the development of an innovative instrument called Network Security Agreements. These examples show that creative administrators are able to design innovative approaches within outdated legal frameworks to at least partially mitigate the outdated laws. Gregory N. Mandel shows how the uncertainties about an emerging technology can be used to leverage, early in the technology's life cycle, all parties to seek consensus on such innovative and flexible approaches.

Anticipation and up-front innovation are only part of the solution, however. As Braden Allenby illustrates with the historical example of railroads, it is impossible to predict the myriad of second- and third-order impacts of a new technology and all their economic, practical and social implications. LeRoy Paddock and Molly Masterton, as well as Mandel in his chapter, set forth a tool box of mechanisms that can be used to make governance more adaptive and responsive to a rapidly evolving technology. Marc A. Saner goes one step further and suggests that we must move beyond governance of a technology and start also thinking of a new paradigm of adaption to new technologies. Jennifer Kuzma draws attention to the importance of considering and involving the public and their opinions in the adaptive response to evolving technologies. Finally, Kenneth W. Abbott in his introductory chapter and Gary E. Marchant and Wendell Wallach in their proposal for technology Coordinating Councils focus on the need for coordination and orchestrating the multiple legal instruments, governance approaches, and stakeholders that will be involved in the oversight of any emerging technology.

Taken together, this book offers a series of diverse but complementary governance models, approaches and proposals for improving the management of emerging technologies. None of the proposals is perfect or sufficient on its own – indeed, many of the suggested approaches would work well in conjunction with other proposals in the book. Even in combination, the resulting improvements in governance from implementing the proposals contained herein are likely to be incremental, imperfect, and tentative. In contrast, however, adhering to the traditional regulatory system alone is certainly inadequate, unacceptable and negligent.

The time for speculating in the abstract about managing emerging technologies is past; it is now time for action. Emerging technologies such as biotechnology, synthetic biology, genomics, nanotechnology, robotics, applied neuroscience, and information technologies are no longer emerging, they are here now and in the present, albeit ever-changing. After percolating in laboratories for the past two to three decades, with their main societal application being science fiction scenarios, they are now entering the commercial marketplace, with real companies, products, applications, benefits and risks. Given the immense challenges in managing the risks of these technologies, the best we can hope for is to "muddle" through.[8] But even achieving that modest goal will require unprecedented innovation, experimentation, and resolve. Hopefully this book has provided some ideas and proposals that can help to steer the governance of emergent technologies in that direction.

---

[8]   Charles E. Lindblom, "The Science of 'Muddling Through,' 19 PUB. ADMIN. REV. 79, 79–80 (1959).

Index